Discovery EDUCATION | SCIENCE **TECHBOOK**

Unit 2
Materials from the Land

Copyright © 2020 by Discovery Education, Inc. All rights reserved. No part of this work may be reproduced, distributed, or transmitted in any form or by any means, or stored in a retrieval or database system, without the prior written permission of Discovery Education, Inc.

NGSS is a registered trademark of Achieve. Neither Achieve nor the lead states and partners that developed the Next Generation Science Standards were involved in the production of this product, and do not endorse it.

To obtain permission(s) or for inquiries, submit a request to:
Discovery Education, Inc.
4350 Congress Street, Suite 700
Charlotte, NC 28209
800-323-9084
Education_Info@DiscoveryEd.com

ISBN 13: 978-1-68220-791-8

Printed in the United States of America.

  6 7 8 9 10 CWM 26 25 24 23   B

Acknowledgments

Acknowledgment is given to photographers, artists, and agents for permission to feature their copyrighted material.

Cover and inside cover art: Byelikova Oksana / Shutterstock.com

Table of Contents

Unit 2
Letter to the Parent/Guardian v

Materials from the Land viii

   Get Started: Shaping the Land with Trash 2

Unit Project Preview: Choosing the Best Materials 4

Concept 2.1
Material Properties 6

   Wonder 8

      Let's Investigate Build a House 10

   Learn 22

   Share 62

Concept 2.2
Changing Materials 70

   Wonder 72

      Let's Investigate A Smart Little Pig 74

   Learn 84

   Share 140

Concept 2.3
Materials in Design 148

Wonder 150
Let's Investigate Three Pigs Problem 152

Learn 164

Share 212

Unit Project
Unit Project: Choosing the Best Materials 224

Grade 2 Resources
Bubble Map R1
Safety in the Science Classroom R3
Vocabulary Flash Cards R7
Glossary R17
Index R30

Dear Parent/Guardian,

This year, your student will be using Science Techbook™, a comprehensive science program developed by the educators and designers at Discovery Education and written to the Next Generation Science Standards (NGSS). The NGSS expect students to act and think like scientists and engineers, to ask questions about the world around them, and to solve real-world problems through the application of critical thinking across the domains of science (Life Science, Earth and Space Science, Physical Science).

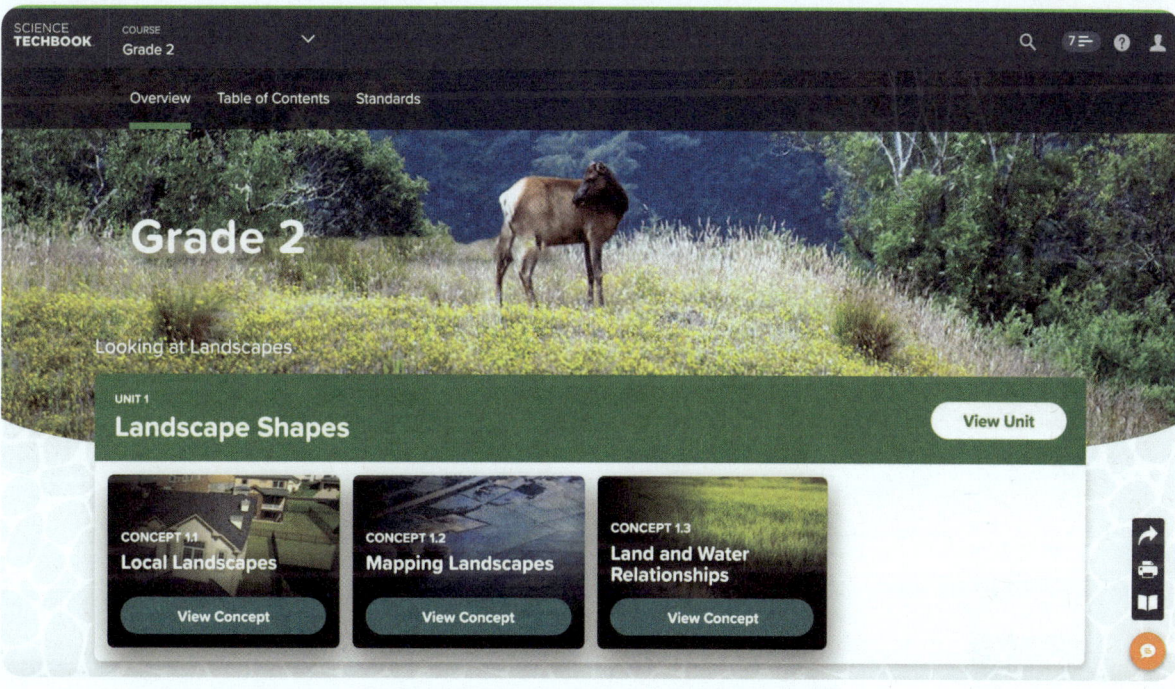

Unit 2: Materials from the Land

Science Techbook is an innovative program that helps your student master key scientific concepts. Students engage with interactive science materials to analyze and interpret data, think critically, solve problems, and make connections across science disciplines. Science Techbook includes dynamic content, videos, digital tools, Hands-On Activities and labs, and gamelike activities that inspire and motivate scientific learning and curiosity.

You and your child can access the resource by signing in to www.discoveryeducation.com. You can view your child's progress in the course by selecting the Assignment button.

Science Techbook is divided into units, and each unit is divided into concepts. Each concept has three sections: Wonder, Learn, and Share.

Units and Concepts Students begin to consider the connections across fields of science to understand, analyze, and describe real-world phenomena.

Wonder Students activate their prior knowledge of a concept's essential ideas and begin making connections to a real-world phenomenon and the **Can You Explain?** question.

Learn Students dive deeper into how real-world science phenomenon works through critical reading of the Core Interactive Text. Students also build their learning through Hands-On Activities and interactives focused on the learning goals.

Share Students share their learning with their teacher and classmates using evidence they have gathered and analyzed during Learn. Students connect their learning with STEM careers and problem-solving skills.

Within this Student Edition, you'll find QR codes and quick codes that take you and your student to a corresponding section of Science Techbook online. To use the QR codes, you'll need to download a free QR reader. Readers are available for phones, tablets, laptops, desktops, and other devices. Most use the device's camera, but there are some that scan documents that are on your screen.

For resources in Science Techbook, you'll need to sign in with your student's username and password the first time you access a QR code. After that, you won't need to sign in again, unless you log out or remain inactive for too long.

We encourage you to support your student in using the print and online interactive materials in Science Techbook on any device. Together, may you and your student enjoy a fantastic year of science!

Sincerely,

The Discovery Education Science Team

Unit 2: Materials from the Land | vii

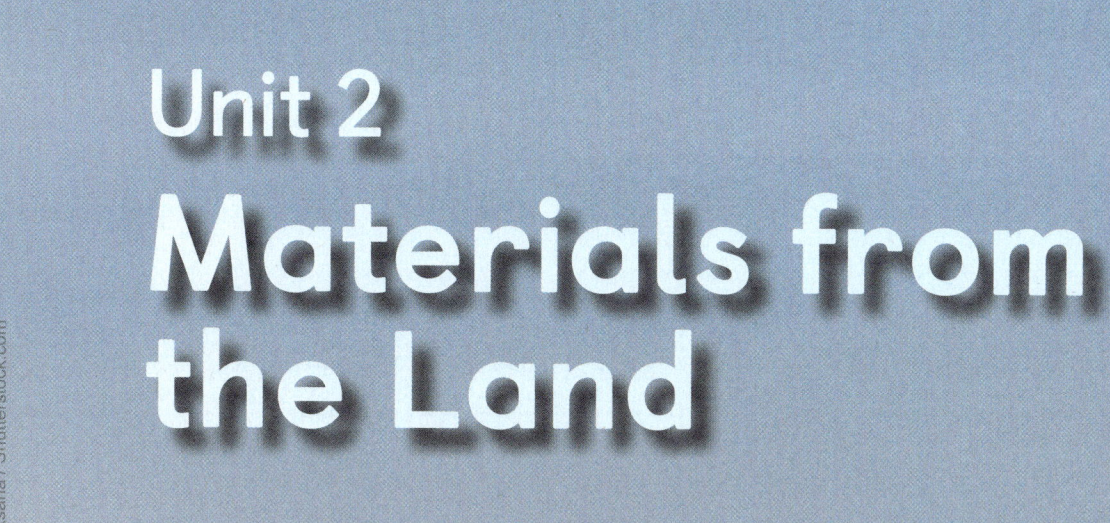

Unit 2
Materials from the Land

Get Started

Shaping the Land with Trash

What happens when you throw away trash? Much of the trash is placed in the ground in a landfill. As more trash is added, the landfill grows into a huge hill. Humans are shaping the landscape. In this unit, you will learn about materials that shape our land. You will learn how different materials are described and what happens when you mix them, heat them, or cool them. In the end, you will use what you know about the properties of materials to design useful, eco-friendly objects.

Watch the video to learn about what happens to trash made of different materials.

Quick Code: us2255s

Shaping the Land with Trash

Think About It

Look at the photograph. **Think** about the following questions:

- How can we describe different materials?
- How are materials similar to and different from one another?
- What sorts of changes can happen to materials?
- How can the materials we use affect the environment?

African Hut

Unit 2: Materials from the Land | 3

Unit Project Preview

 **Design Solutions Like a Scientist**

Quick Code: us2256s

Hands-On Engineering: Choosing the Best Materials

In this activity, you will take apart an object. You will observe how its parts work together. Then you will design an object that performs a similar function using recycled or reused materials.

Organized Recycling Bin

CCC Patterns

SEP Constructing Explanations and Designing Solutions

CCC Structure and Function

Ask Questions About the Problem

You are going to take apart an object. You will observe how its parts work together. Then you will design an object that performs a similar function using recycled or reused materials. **Write** some questions you can ask to learn more about the problem. As you work on different activities in the unit, **write** down answers to your questions.

Unit 2: Materials from the Land

CONCEPT
2.1

Material Properties

Student Objectives

By the end of this lesson:

- [ ] I can observe patterns in different materials found in the landscape.

- [ ] I can collect information and analyze materials to identify their properties.

- [ ] I can plan and conduct an investigation to compare, sort, and describe materials based on their properties.

Key Vocabulary

- [ ] absorb
- [ ] engineer
- [ ] flexibility
- [ ] hardness
- [ ] material
- [ ] property
- [ ] soil

Quick Code: us2258s

2.1: Material Properties | 7

Activity 1
Can You Explain?

How are materials alike and different?

_____
_____
_____
_____
_____

Quick Code:
us2259s

2.1: Material Properties

2.1 | Wonder How are materials alike and different?

 **Activity 2**
Ask Questions Like a Scientist

Quick Code:
us2260s

Build a House

Watch the video. Pay attention to the **materials** that the pigs use to build their houses. Then, **talk** about the question.

Let's Investigate Build a House

SEP Planning and Carrying Out Investigations

 **Talk Together**

What kinds of **materials** did the pigs use? Talk about whether they worked well for building the houses.

Imagine you could talk to the pigs about the choices they made to build their houses. What questions would you ask about the materials?

_____

_____

Draw a picture of a house. **Label** the materials you would use to build it.

Your Drawing

2.1: Material Properties | 11

Activity 3
Analyze Like a Scientist

Building a House

Read the text. Then, **underline** reasons why it's important to analyze soil before building a house.

Quick Code: us2261s

> Read Together

Building a House

When you think about the materials to build a house, do you think about **soil**? Soil is very important when building a house. If the soil under the house is too loose, the house will collapse. Shifting layers of soil can make a building crack. Wet soil underneath a house can also cause damage. Soil science engineers search the landscape for just the right soil. They help builders find good soil so that the houses they construct will last.

Soil

2.1: Material Properties | 13

2.1 | Wonder — How are materials alike and different?

Activity 4
Investigate Like a Scientist

Quick Code: us2262s

Hands-On Investigation: What's in That Soil?

In this activity, you will use a hand lens or magnifying glass to observe soil. You will then describe what you see.

Make a Prediction

You are going to observe three different kinds of soil. **Write** or **draw** your predictions.

How do you think the soils will look the same? How will they look different?

SEP Planning and Carrying Out Investigations

What materials do you need? (per group)

- Disposable trays
- Disposable gloves
- Hand lens
- Plastic cup, 9 oz
- Soil, sandy
- Soil, clay
- Soil, potting
- Soil, compost

What Will You Do?

Look at one kind of soil using the materials from your teacher. **Pour** the soil onto the tray. **Look** at the soil with a hand lens or magnifying glass. **Write** or **draw** what you see.

2.1: Material Properties

2.1 | Wonder   How are materials alike and different?

Look at another kind of soil. **Pour** the soil onto the tray in a different place than the first soil. **Look** at the soil with a hand lens or magnifying glass. **Write** or **draw** what you see.

Look at the last kind of soil. **Pour** this soil onto the tray in a different place than the other soils. **Look** at the soil with a hand lens or magnifying glass. **Write** or **draw** what you see.

Think About the Activity

Pick two of the soils you observed.

How are the two soils alike?

How are the two soils different?

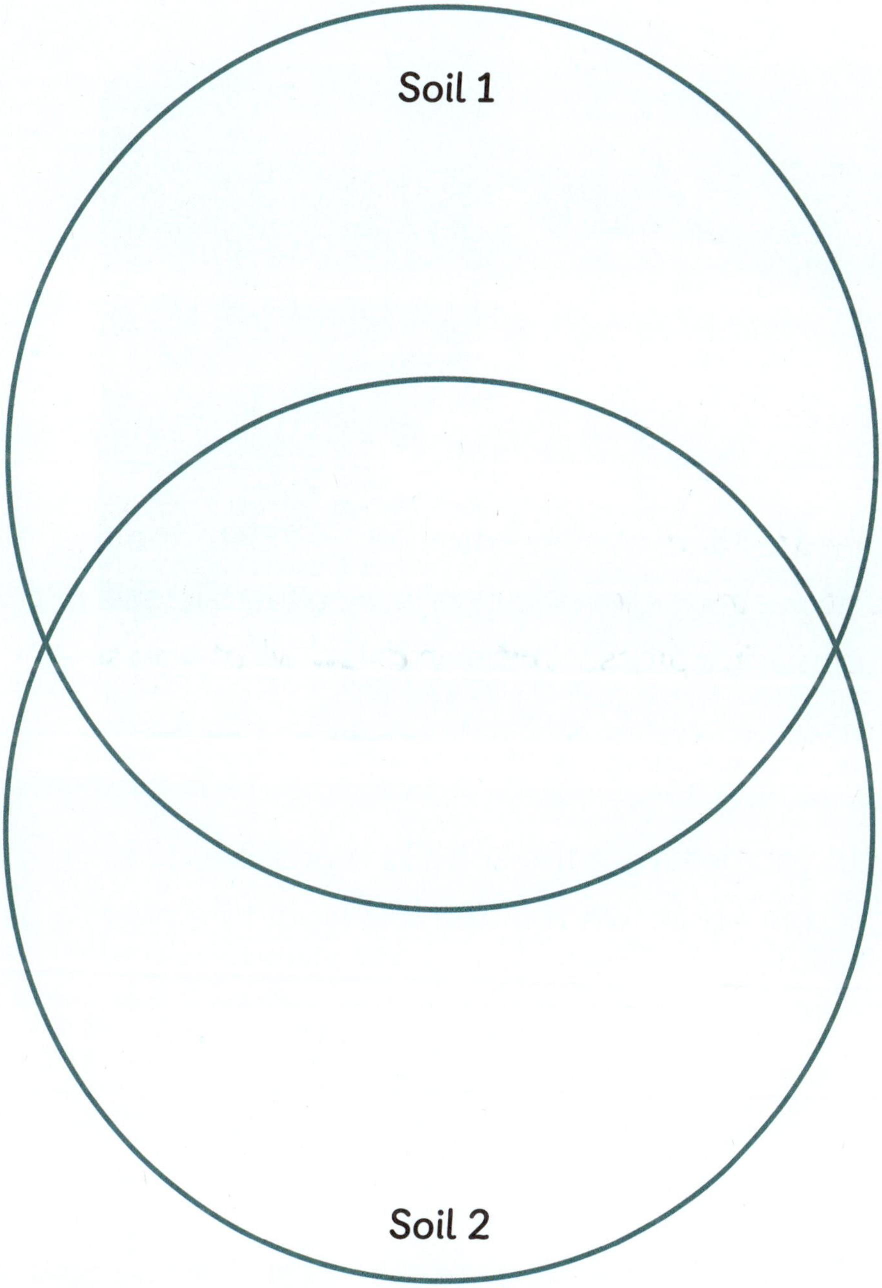

Soil 1

Soil 2

2.1: Material Properties | 17

2.1 | Wonder — How are materials alike and different?

Activity 5
Observe Like a Scientist

Quick Code:
us2263s

Beach Path

Look at the photo.

Beach Path

How would you describe the soil? Is it possible to build a house on this landscape? If so, what materials would be best to use? Why?

_____
_____
_____
_____

Think about what you might see walking on a sandy beach path. Then **write** or **draw** about it.

I see...	I think...	I wonder...

2.1: Material Properties | 19

2.1 | Wonder How are materials alike and different?

Activity 6
Evaluate Like a Scientist

Quick Code:
us2264s

What Do You Already Know About Material Properties?

Compare the Toys

Look at the photos. **Circle** the words that describe each toy. Words can be used more than once.

Marbles

(hard) (round) (small) smooth (red) medium fuzzy (yellow)

Tennis Ball

(hard) (round) small smooth red (medium) (fuzzy) (yellow)

What Objects Are Made Of

Look at a bottle of glue. **Write** as many words as you can to describe the bottle of glue.

I can see orange blue a top white words QR code black dark blue.

Draw what you see.

Your Drawing

[Drawing of a bottle labeled "Eleners Glue" with "safe non toxic"]

2.1: Material Properties 21

2.1 | Learn How are materials alike and different?

What Materials Can Be Found Where You Live?

Activity 7
Observe Like a Scientist

Quick Code: us2265s

Schoolyard Landscape

Look at the photo.

Schoolyard Landscape

22

What different materials do you see in this landscape?

plastic wood metal
dirt grass paint
Glass concrete

2.1: Material Properties | 23

2.1 | Learn How are materials alike and different?

Activity 8
Think Like a Scientist

Quick Code: us2266s

Materials in the Schoolyard Landscape

In this activity, **use** your senses to observe materials where you live. **Describe** the materials you find and how they are used.

What materials do you need? (per group)
- Pencils
- Colored pencils

What materials do you think you will find? How will the materials be used?

SEP Planning and Carrying Out Investigations   **CCC** Patterns

What Will You Do?

Look for different materials in the schoolyard. **Write** or **draw** what you see.

Material	Words to Describe It	What Is It Used For?

2.1 | Learn   How are materials alike and different?

Think About the Activity

What different materials did you find?

Material	Words to Describe Material	Objects Made with Material
Metal		
Wood		
Plastic		
Concrete		

Activity 9
Observe Like a Scientist

Quick Code:
us2267s

Transportation

Watch the video. Pay attention to the steps for making asphalt and steel.

Transportation

Now, talk together about the steps involved in making asphalt and steel. Describe how the material for roads is made.

2.1: Material Properties | 27

2.1 | Learn   How are materials alike and different?

Now, **read** the text. The text repeats what was said in the video. Look for details about the process of making either asphalt or steel.

Minerals are used to make almost everything that is used for transportation as well, even the roads we drive on. First, aggregate is mixed with an oil product called "liquid asphalt" in this rotating heating drum. Roadways are covered with a special sharp rock called "asphalt base." The asphalt is then dumped out over the base, flattened, smoothed, and left to harden. Then our cars can drive on it.

Steel is an important part of every one of these vehicles, and steel is made from iron ore. Ore is a type of rock that has a lot of one metal. In this case, it's iron. The iron ore is mixed with coal, limestone, and other minerals. When it is heated, it changes. Then the red-hot steel can be shaped and molded into many kinds of vehicles. Steel is so strong it can carry oil all the way across Alaska or support a bridge across the bay. It even holds up huge skyscrapers.

Choose either the asphalt or steel process and **draw** and **label** pictures that show each step in the process.

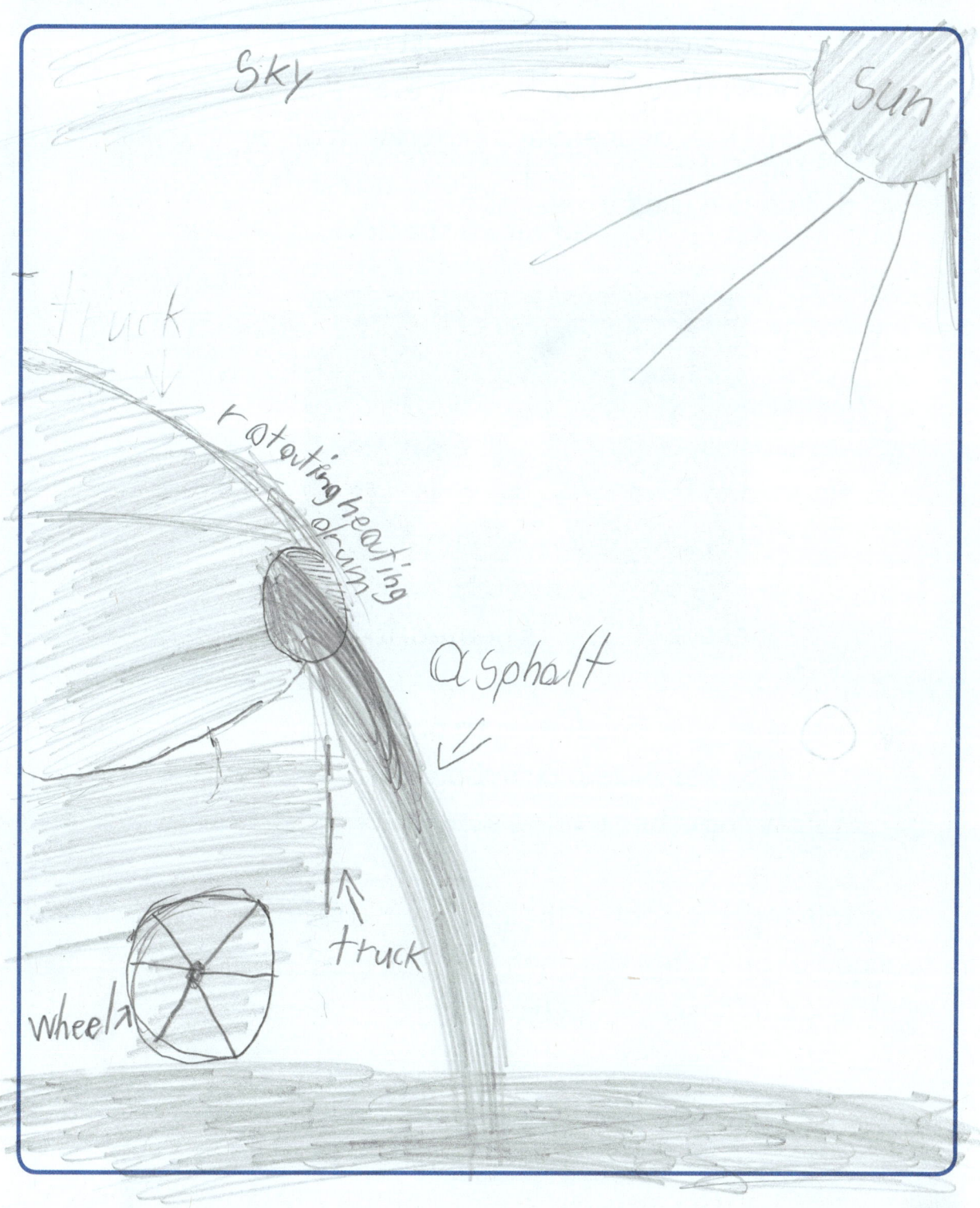

2.1: Material Properties

2.1 | Learn — How are materials alike and different?

Activity 10
Observe Like a Scientist

Quick Code: us2268s

Making Observations: Soil

Watch the video about different kinds of soil. Pay attention to the words that are used to describe soil.

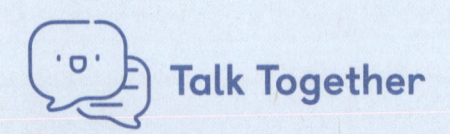

Making Observations: Soil

Talk Together

Now, talk together about the different kinds of soils you learned about. What did each soil look like? What words were used to describe each soil?

What is soil made of?

Rock sand minerals clay

Choose one kind of soil—(sand), clay, or loam soil. **Write** or **draw** to describe the soil.

pretty soft

2.1: Material Properties | 31

Activity 11
Analyze Like a Scientist

Quick Code: us2269s

Different Landscape Materials

Read the text. **Underline** a material that would be useful for building a house because it's hard and sturdy.

Read Together

Different Landscape Materials

Look outside at the landscape of your school. What do you see? Do you see grass-covered hills? Do you see areas with brown soil? Do you see a concrete sidewalk? Do you see a play area with a rubber mat? There are many different materials in a landscape.

Some materials in a landscape are natural, like soil, rocks, and <u>wood</u>. Other materials are made by humans, like <u>concrete</u> or asphalt.

SEP Engaging in Argument from Evidence

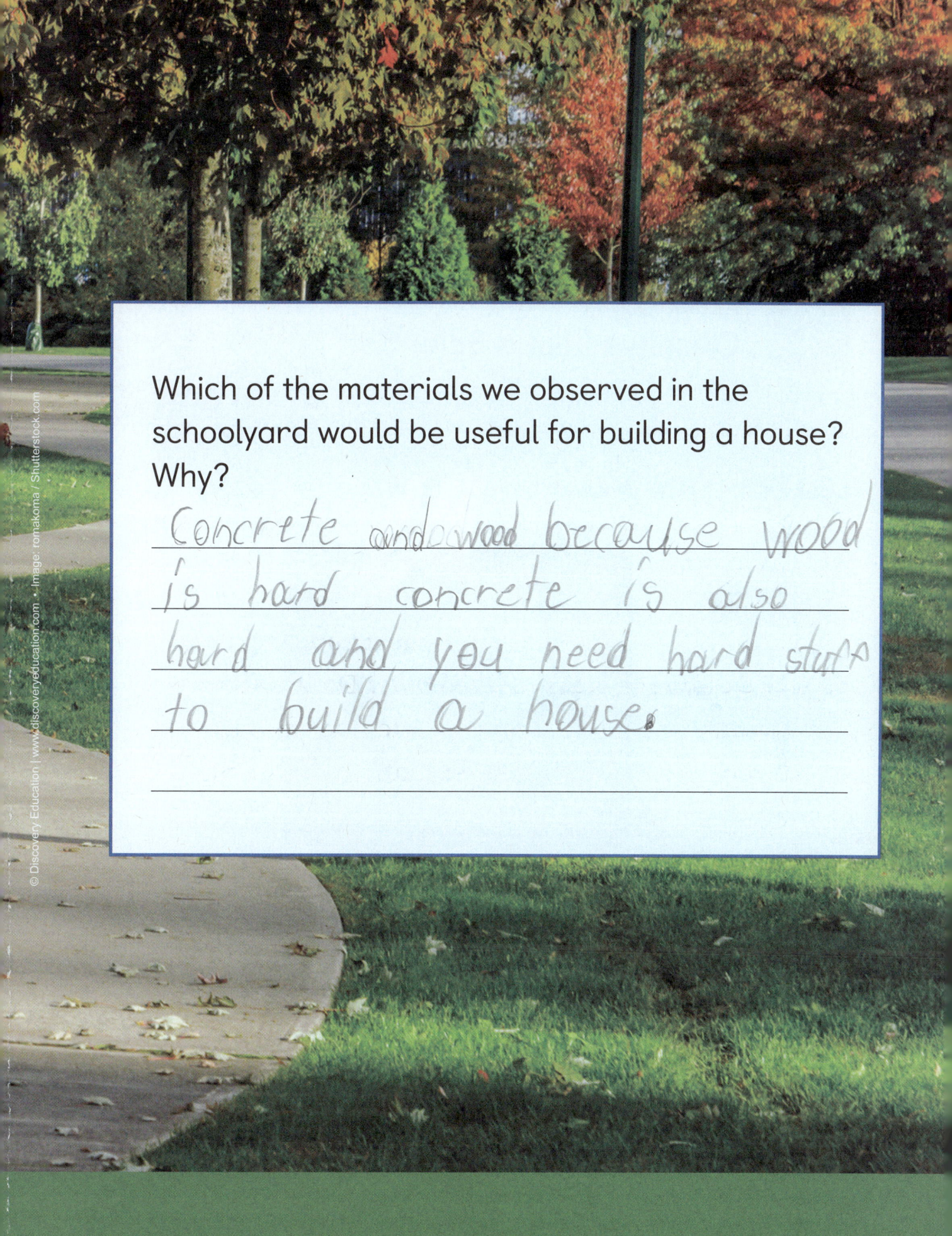

Which of the materials we observed in the schoolyard would be useful for building a house? Why?

Concrete and wood because wood is hard concrete is also hard and you need hard stuff to build a house.

2.1: Material Properties | 33

2.1 | Learn How are materials alike and different?

How Are Materials Described?

Activity 12
Observe Like a Scientist

Quick Code: us2270s

Chalk Rock

Look at the photo. **Answer** the questions.

Chalk Rock

How would you describe this rock? Do you have a rock like this in your classroom?

I do have this rock in my classroom but I have used it, it feels rough but, it's like chalk.

SEP Constructing Explanations and Designing Solutions

Observe a chalk rock. How would you describe it? Is it bumpy or smooth? Is it hard or soft?

_____

_____

Observe another object in the classroom. How would you describe this object? **Draw** and **write** about it.

2.1: Material Properties | 35

Activity 13
Analyze Like a Scientist

Quick Code: us2271s

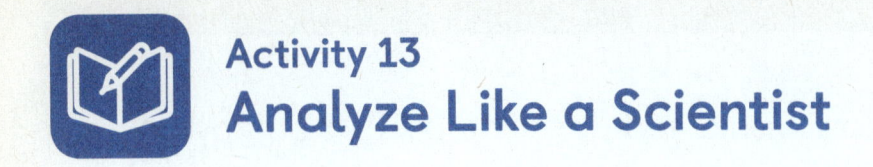

All About Objects

Read the text. **Underline** ways to describe objects.

Read Together

All About Objects

We can describe objects in many ways. Objects may hold a little or a lot of water. These objects **absorb** things. This is called absorption. Objects may bend easily or be rigid and not bend. This is called **flexibility**. Objects may be easily broken or not. This is called **hardness**. Something you can observe about an object is called a **property**.

Gallon of Milk

Objects can be a solid, liquid, or gas. A solid holds its own shape. A rock is a solid. A liquid can flow. It takes the shape of the container it is in. Milk and juice are liquids. An empty box is filled with air. Air is a gas. The air takes the shape of the box.

SEP Engaging in Argument from Evidence

36

Activity 14
Observe Like a Scientist

Quick Code: us2272s

Rocks Are Different

Watch the video. **Look** for differences in rocks.

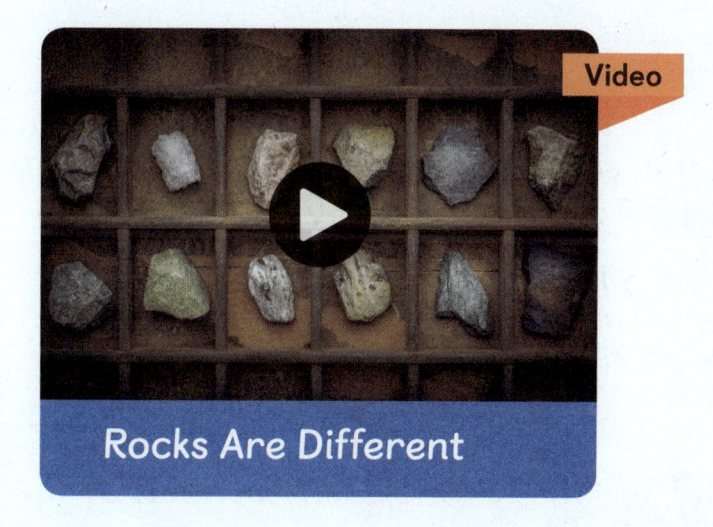

Video

Rocks Are Different

Talk Together

Now, talk together about the different properties of rocks, such as shape, color, hardness, and texture.

2.1: Material Properties

2.1 | Learn How are materials alike and different?

Activity 15
Investigate Like a Scientist

Quick Code: us2273s

Hands-On Investigation: Rock Classification

In this activity, you will observe rocks. Then you will sort them.

Make a Prediction

You are going to sort rocks by shape, color, and texture. **Write** or **draw** your predictions.

What properties do you think the rocks will have that are the same?

> smoth   hard   soft shiny

What properties do you think the rocks will have that are different?

> big   small   color

SEP Planning and Carrying Out Investigations   **CCC** Patterns

What materials do you need? (per group)

- Hand lens
- Quartz milky
- Granite rock, medium
- Sandstone, red
- Conglomerate rock, medium
- Slate, gray
- Gneiss rock, medium
- Safety goggles (per student)

HANDS-ON INVESTIGATION

What Will You Do?

Look at the rocks with a hand lens. **Write** or **draw** what you see. **Describe** their color and size. **Describe** their shape. What do they feel like? **Tell** how you sorted the rocks.

Some rock felt soft shiny hard brown gray with black clear chalky

2.1: Material Properties | 39

2.1 | Learn   How are materials alike and different?

Think About the Activity

How did you sort the rocks?

[student drawing: "soft" with two rocks, "hard" with three rocks]

How did you use your senses to sort the rocks by shape?
Write and **draw**.

My hands feel super smoth.

[student drawing of a hand]

How did you use your senses to sort the rocks by color?
Write and **draw**.

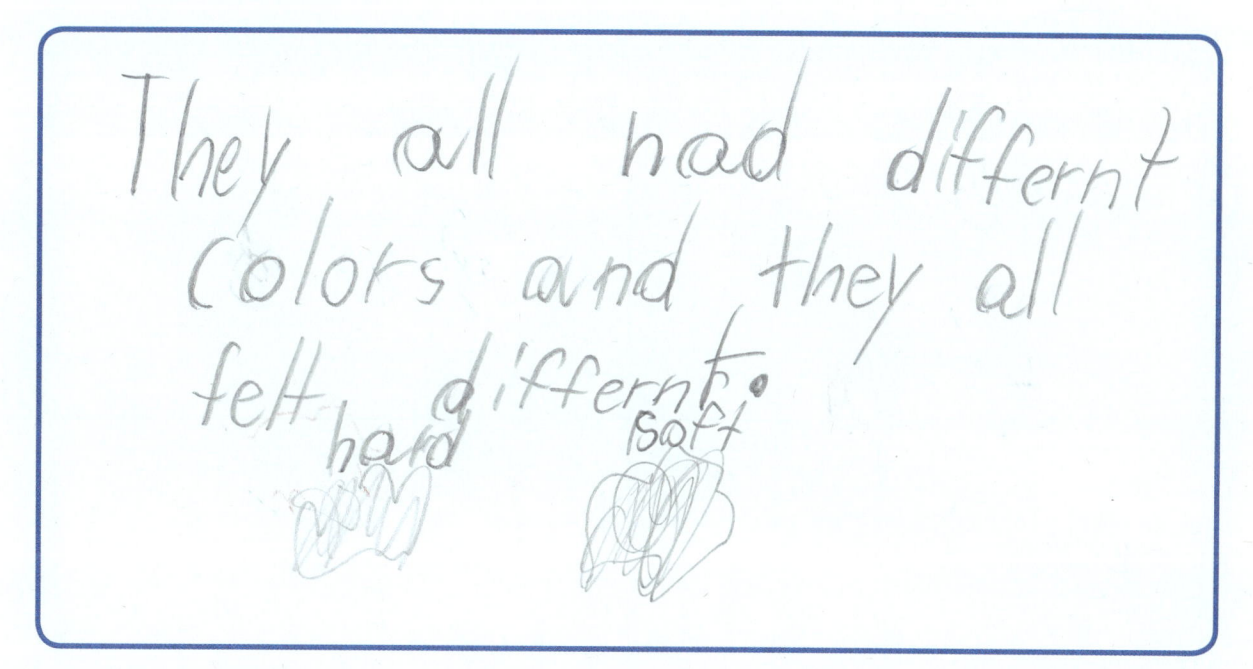

> They all had differnt colors and they all felt differnt.
> hard   soft

How did you use your senses to sort the rocks by texture?
Write and **draw**.

> Some rocks were rough small and big.

2.1: Material Properties | 41

2.1 | Learn   How are materials alike and different?

What patterns did you see in the rocks? **Write** about them. **Draw** pictures.

Patterns	Drawing
bumpy hard soft bumpy hard soft	bumpy, hard, soft, bumpy, hard, soft

Think about the rocks you looked at. What questions do you have about the rocks? **Write** them.

_____
_____
_____

Activity 16
Observe Like a Scientist

Quick Code: us2274s

Sorting Objects: Investigating and Classifying Materials

Watch the video. **Listen** for the different properties that are used to describe classroom objects. Then, **talk** about the question. Finally, **compare** two classroom objects.

Video

Sorting Objects: Investigating and Classifying Materials

Talk Together

Now, talk together about how the properties of objects are alike and different.

2.1: Material Properties

2.1 | Learn How are materials alike and different?

Choose two objects from the classroom. **Observe** the objects. **Compare** their properties.

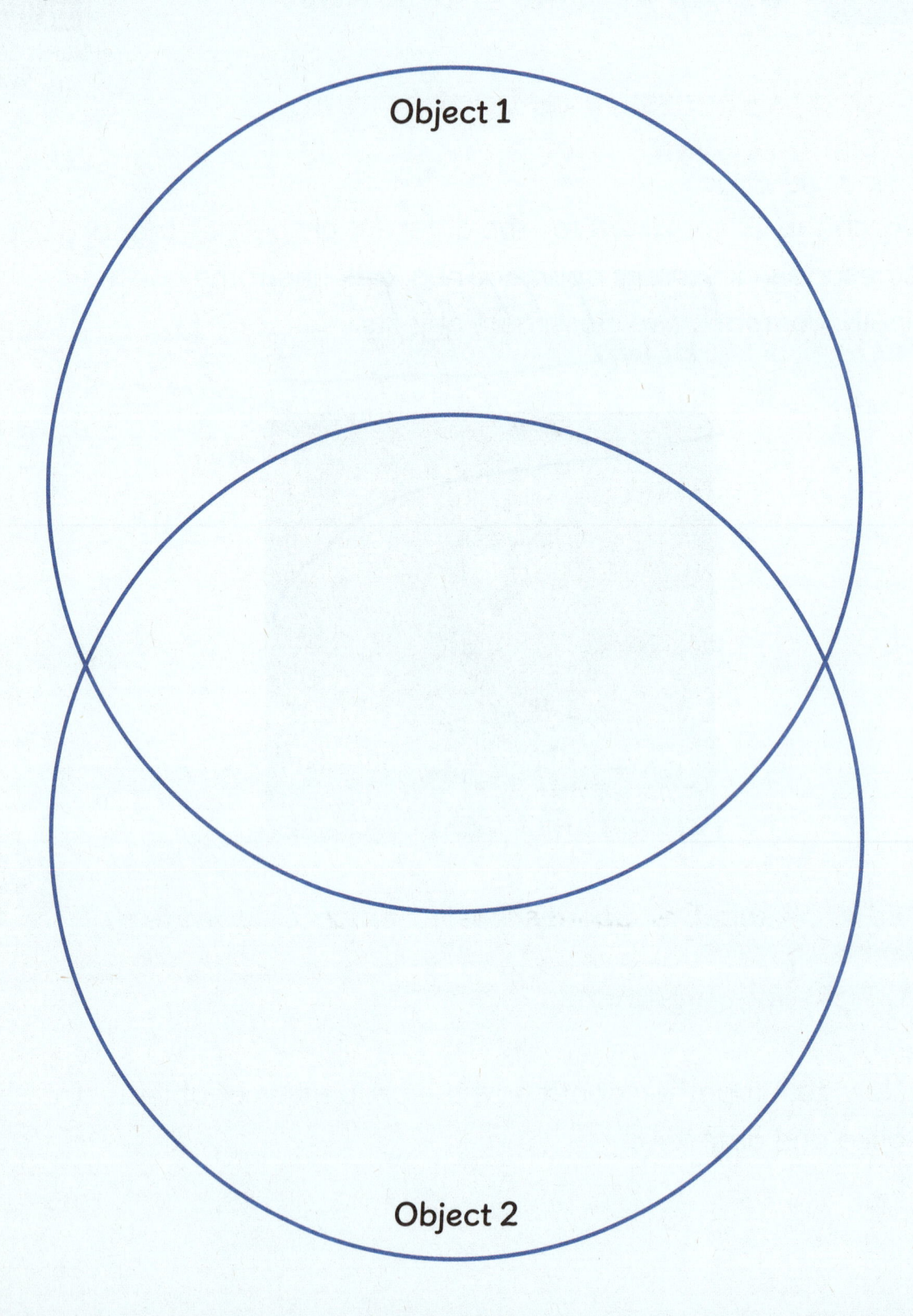

Object 1

Object 2

Activity 17
Observe Like a Scientist

Quick Code: us2275s

Music Video: It's a Property

Watch the video.

Music Video: It's a Property

Talk Together

Now, talk together about some good dance moves for this song.

2.1: Material Properties | 45

2.1 | Learn — How are materials alike and different?

Activity 18
Observe Like a Scientist

Quick Code: us2276s

Water in a Landscape

Look at the photo.

Water in a Landscape

SEP Engaging in Argument From Evidence

46

Answer the questions.

How would you describe the water in this landscape?

_____
_____
_____
_____

What evidence do you observe in this picture that water can exist in three different forms?

_____
_____
_____
_____

2.1 | Learn How are materials alike and different?

Activity 19
Evaluate Like a Scientist

Quick Code: us2277s

What Is My State?

Look at the picture. Can you find things that are solids, liquids, or gases?

Rocks Jut from Landscape

CCC Patterns

48

Look closer at the objects found in the picture. **Write** the correct word under each picture to tell if it is a solid, a liquid, or a gas.

solid

liquid

gas

solid

2.1: Material Properties

Read Together

Comparing Materials

In the Three Little Pigs story, the first two pigs built their houses using straw and sticks. The pigs could have thought more about these materials. They might have chosen different materials if they knew a strong wind could destroy their work. Comparing a variety of materials can be helpful. Do you think the pigs' houses would have kept them safe in the rain?

Engineers compare materials. They observe materials and test their properties. First, they decide what property of an object they want to test. For example, the engineers might want to know which material is the hardest or which material is waterproof. Then, they observe what happens when each different material is tested in the same way.

Engineers use materials with just the right properties in their designs. They test products. Engineers use product testing to make sure an object has the properties they want. The product is ready when it meets a given criterion or standard. A criterion is a quality that is necessary for a proposed design to be considered a good design.

How Are Materials Compared?

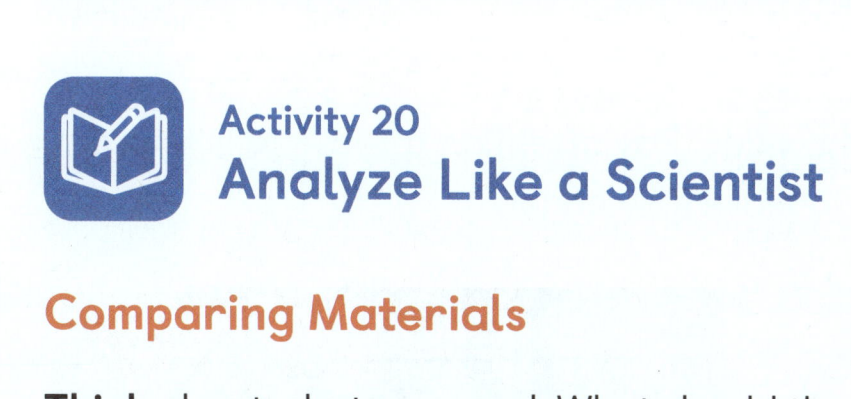

Activity 20
Analyze Like a Scientist

Quick Code: us2278s

Comparing Materials

Think about what you read. What should the little pigs have done before choosing materials for their houses?

They should have tested the houses before.

Draw what you think they should have done.

[drawing of a brick house labeled "brick house"]

2.1: Material Properties | 51

2.1 | Learn How are materials alike and different?

Activity 21
Observe Like a Scientist

Quick Code: us2279s

Investigating House Materials

Watch the video. Pay attention to ways to improve the design of a house to make it stronger.

Investigating House Materials

Write a paragraph to help the pigs build a better house.
Write three sentences to tell them what to do.

Little pig, take my advice.

You should build a house out of blocks because it can stand rain and wind.

Your house will be better if you follow my advice.

Activity 22
Observe Like a Scientist

Quick Code: us2280s

What Is Product Testing?

Watch the video. **Look** for the properties that different products are tested for.

What Is Product Testing?

Talk Together

Now, talk together about why product testing is important.

2.1: Material Properties

2.1 | Learn — How are materials alike and different?

Activity 23
Investigate Like a Scientist

Quick Code: us2281s

Hands-On Investigation: Comparing Different Materials

In this activity, you will use your senses to observe different materials. You will compare objects that have the same purpose but are made of different materials. Then, you will record your observations.

Make a Prediction

You are going to observe properties of objects made with different materials. **Write** or **draw** your predictions.

Why do you think it is important for objects to be strong or hard?

> I think they should be strong so they can doust a lot.

Why do you think it is important for objects to be absorbent?

> I think it's important for objects to be absorbent because when I spill a liquid I can clean it up easly.

What materials do you need? (per group)

- Paper shopping bag
- Plastic shopping bag
- Cloth shopping bag
- Plastic milk container
- Plastic shovel
- Metal shovel
- Plastic toy building block
- Wood block
- Balance, double pan
- Graduated cylinder, 50 × 1 mL
- Hand lens
- Hand towel
- Metric ruler
- Paper towels

HANDS-ON INVESTIGATION

What Will You Do?

Observe the objects at each station. What are they made of? Which is the strongest? Which is the best for the job they are meant to do? **Write** and **draw**.

Station 1

The objects are made of plastic and paper. The strongest is the plastic bag.

SEP Planning and Carrying Out Investigations　　**CCC** Patterns

2.1: Material Properties

2.1 | Learn   How are materials alike and different?

What are the bags made of?	How many units did it take to break the bag?

Station 2

What are the towels made of?	How much water did each type of towel absorb?

Station 3

What are the shovels made of?	How many rocks did it take to bend or break the shovels?

2.1: Material Properties

2.1 | Learn How are materials alike and different?

Station 4

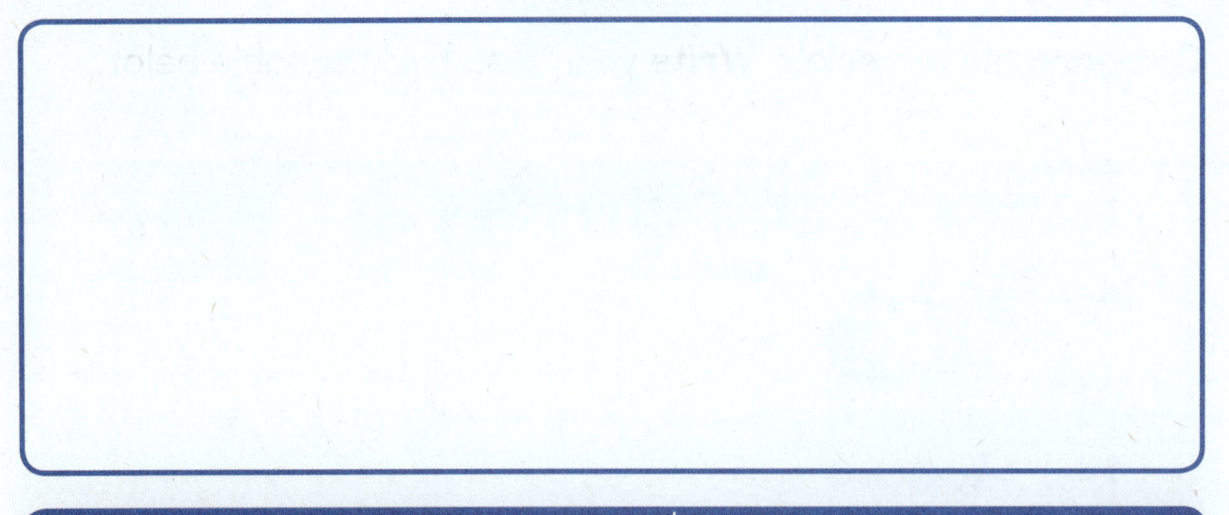

What are the blocks made of?	At what height did the block or rock bend or chip?

Measure each of the materials in some way and **record** these measurements in the tables.

Think About the Activity

Compare the materials. **Write** your results in the table below.

Station Number	1	2	3	4
Name of the object				
What are the objects made of?				
Which object is stronger/absorbs more?				
Which object is better for the job?				

2.1: Material Properties | 59

2.1 | Learn — How are materials alike and different?

Graph your results.

9								
8								
7								
6								
5								
4								
3								
2								
1								

Activity 24
Evaluate Like a Scientist

Quick Code: us2282s

What's the Matter?

Matter is all around you. How can we describe it? **Draw** lines to **match** each picture with the description that tells about it.

This is smooth, blue, and wet. It is a liquid.

This is rough, brown, and heavy. It is a solid.

This is cold, smooth, and hard. It is a solid.

This is pink and yellow. It can bend. It is a solid.

SEP Engaging in Argument From Evidence

2.1: Material Properties

2.1 | Share — How are materials alike and different?

Activity 25
Record Evidence Like a Scientist

Quick Code: us2283s

Build a House

Now that you have learned about material properties, look again at the video Build a House. You first saw this in Wonder.

Let's Investigate Build a House

Talk Together

How can you describe Build a House now? How is your explanation different from before?

SEP Constructing Explanations and Designing Solutions

Look at the Can You Explain? question. You first read this question at the beginning of the lesson.

> **Can You Explain?**
>
> How are materials alike and different?

Now, you will use your new ideas about Build a House to answer a question.

1. **Choose** a question. You can use the Can You Explain? question, or one of your own. You can also use one of the questions that you wrote at the beginning of the lesson.

My Question

how are materials alike and diffrent??

2. Then, **use** the sentence starters on the next page to **answer** the question.

2.1: Material Properties | 63

2.1 | Share — How are materials alike and different?

I know that materials are alike and different because
they turn into _____ are all solid they are different ____ ____ and that they all have different properties.

Based on my observations, I think
houses should be made of cement

would be best for building houses because

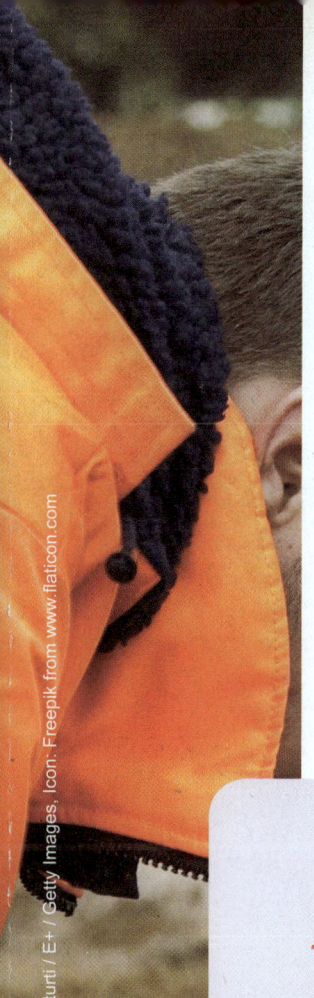

STEM in Action

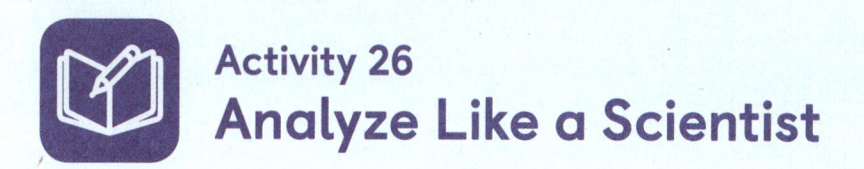

Activity 26
Analyze Like a Scientist

Quick Code: us2284s

A Dirty Job

Read the text about soil scientists. **Look** for details about the kinds of soil that are good for building.

> **Read Together**

A Dirty Job

Do you like playing in dirt? Some scientists study dirt for a job! They know how soil is made. They know how plants and animals live in soil. They help farmers decide what plants to grow. Some soil scientists even study layers of soil. Soil layers help them find out about the past.

SEP Obtaining, Evaluating, and Communicating Information

2.1: Material Properties | 65

Read Together

Good soil has humus, clay, silt, and sand all jumbled together. Soil starts with rock that gets broken down through chemical and mechanical weathering. Then as plants and animals die, they decompose into well-rotted organic matter we call humus. In the end, soil has a mixture of things. The best soil has the right balance to the mix and is certainly more than just dirt!

Did you know that some soil scientists are engineers? They help builders construct buildings. If the soil under a building is too loose, the building will fall. Shifting layers of soil will crack the building. Soil science engineers help builders find good soil. They help make buildings last.

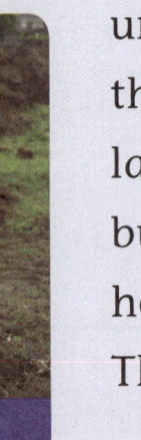

A Soil Scientist at Work

Helping the Builder

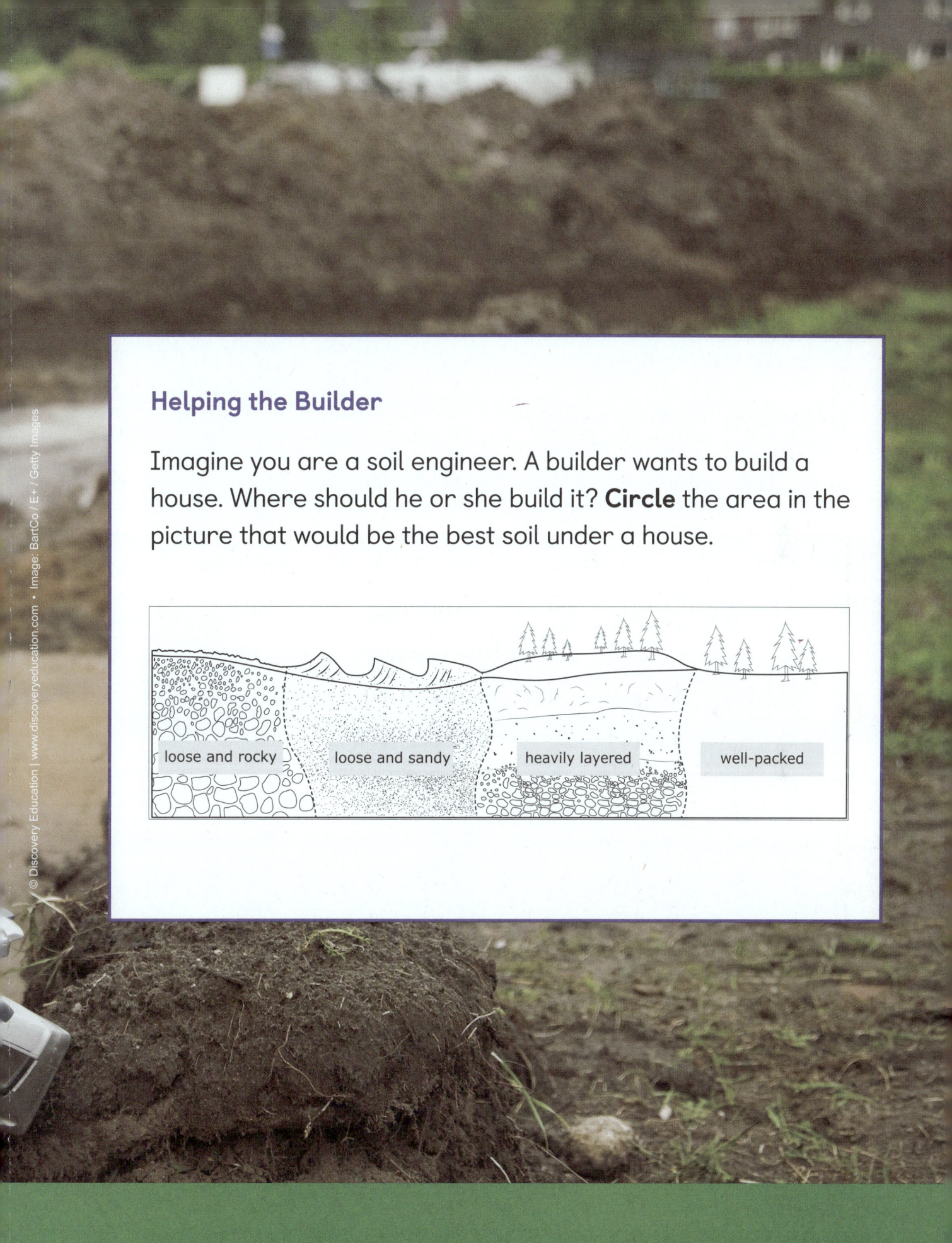

Imagine you are a soil engineer. A builder wants to build a house. Where should he or she build it? **Circle** the area in the picture that would be the best soil under a house.

| loose and rocky | loose and sandy | heavily layered | well-packed |

2.1: Material Properties

2.1 | Share — How are materials alike and different?

Activity 27
Evaluate Like a Scientist

Quick Code: us2285s

Review: Material Properties

Think about what you have read and seen.

What did you learn?

Draw what you have learned.

Then, **tell** someone else about what you learned.

`CCC` Patterns

Talk Together

Think about what you saw in Get Started. Use your new ideas to discuss how materials are alike and different.

2.1: Material Properties

CONCEPT 2.2

Changing Materials

Student Objectives

By the end of this lesson:

- ☐ I can use observations to identify patterns when I mix materials together.

- ☐ I can explain how to separate mixtures.

- ☐ I can use evidence to describe the effects of different temperatures on materials.

- ☐ I can use observations to classify material changes as reversible or irreversible.

Key Vocabulary

- ☐ dissolve
- ☐ solution
- ☐ mixture

Quick Code: us2287s

2.2: Changing Materials

Activity 1
Can You Explain?

What changes to objects can be reversed, and what changes cannot?

to put a block on a rock it be reversed breaking a crayon cannot be reversed.

Quick Code: us2288s

2.2: Changing Materials | 73

2.2 | **Wonder** What changes to objects can be reversed, and what changes cannot?

Activity 2
Ask Questions Like a Scientist

Quick Code: us2289s

A Smart Little Pig

Listen to your teacher read the passage. Pay attention to the challenge that the pigs had.

Read Together

A Smart Little Pig

There were once three little pigs.

They were scared of a big, bad wolf.

They built houses to hide in.

Let's Investigate A Smart Little Pig

SEP Analyzing and Interpreting Data

The first little pig used straw.

Straw is not strong.

The wolf blew the straw house down.

The second little pig used twigs.

Twigs are also not strong.

The wolf blew the twig house down.

2.2: Changing Materials

2.2 | **Wonder** What changes to objects can be reversed, and what changes cannot?

Read Together

The third little pig listened in science class. He knew to use more materials.

This little pig made a frame of wood and nails. He made four walls of bricks. He used cement to hold the bricks together.

He put in glass windows with metal locks.

The wolf could not blow down the house! The bricks stayed together.

Now the three little pigs lived together in the strong, sturdy house.

Talk Together

Now, talk together and decide what question you would ask to each pig about the materials they chose.

How did the pigs react to the challenge of being afraid of the wolf?

They built a house.

What do you know about the properties of straw, sticks, and bricks?

straws are weak but sticks are easy to blow but the bricks can stay togeter

What makes bricks so strong?

The pig use cement.

2.2: Changing Materials | 77

2.2 | Wonder
What changes to objects can be reversed, and what changes cannot?

Activity 3
Observe Like a Scientist

Quick Code: us2290s

Making Bricks

Look at the picture. What clues can you **observe** from the picture about how bricks are made?

Making Bricks

78

I see ...	I think ...	I wonder ...

2.2: Changing Materials | 79

2.2 | Wonder
What changes to objects can be reversed, and what changes cannot?

Activity 4
Observe Like a Scientist

Quick Code: us2291s

Brick Building

Look at the picture. Then **answer** the questions.

Brick Building

Do you think the vines on the building will change the bricks? How?

_____

_____

_____

80

Activity 5
Analyze Like a Scientist

Quick Code: us2292s

The Right Material

Read the text. **Answer** the last question and **discuss** it with a partner.

Read Together

The Right Material

Do you remember the story of the Three Little Pigs? The pigs use different materials to build their houses. The first little pig uses straw. The second little pig uses sticks. The third little pig uses bricks.

The wolf in the Smart Little Pig story tried to blow down the brick house. The wind did not change the brick. Have you ever thought about what might have changed the brick? What might have happened if the wolf pounded the brick with a hammer? What might have happened if the wolf mixed water with the brick by spraying the house with a hose? What do you think the wolf could have done to change the bricks?

What can be done to change a brick?

2.2: Changing Materials

2.2 | **Wonder** What changes to objects can be reversed, and what changes cannot?

Activity 6
Evaluate Like a Scientist

Quick Code: us2293s

What Do You Already Know About Changing Materials?

Things that Change

Look at the pictures. Each picture on the top shows an object. Each picture on the bottom shows a change to the object. **Draw** lines to match each picture on the top with its change.

82 | Discovery EDUCATION

What Changed?

These objects all changed. What happened to make them change? How did they change? Next to each object, **write** a word to tell what changed. Write **Shape**, **Color**, or **Temperature**.

folding a newspaper _____

melting ice pop _____

drawing a picture _____

baking bread _____

making a bow _____

painting a wall _____

Discussing Mixtures

Can you think of some things you can mix together? Can you take them apart? **Write** to tell your ideas.

_____

_____

2.2: Changing Materials

2.2 | Learn — What changes to objects can be reversed, and what changes cannot?

What Happens When Materials Are Mixed Together?

Activity 7
Observe Like a Scientist

Quick Code: us2294s

River Pebbles

Look at the photo. Then **answer** the questions.

River Pebbles

These river pebbles have been combined. Do you think the pebbles can be separated? If so, how?

_____
_____
_____
_____
_____

What other objects can be mixed together?

_____
_____

Activity 8
Investigate Like a Scientist

Quick Code: us2295s

Hands-On Investigation: Mixing Up Fun!

In this activity, you will practice making different **mixtures**.

Make a Prediction

What would happen if you mix solids with solids, liquids with liquids, and solids with liquids? **Write** or **draw** your predictions.

Predictions		
solids with solids	liquids with liquids	solids with liquids

SEP Planning and Carrying Out Investigations

CCC Patterns

2.2: Changing Materials | 85

2.2 | Learn   What changes to objects can be reversed, and what changes cannot?

What materials do you need? (per group)

- Plastic cup, 9 oz
- Plastic cup, 5 oz
- Powdered fruit drink mix (brightly colored)
- Bowl, foam
- Scoop of sprinkles in plastic bowl
- Bowls, plastic
- Spoon, plastic
- Paper
- Markers
- Ice cream
- Water

What Will You Do?

Work together to practice mixing liquids and liquids. Pour the half-filled cup of water into the half-filled cup of fruit juice. Mix it with a spoon. **Write** or **draw** what you see.

Work together to practice mixing solids and liquids. Pour the powdered fruit drink mix into the full cup of water. Mix it with a spoon. **Write** or **draw** what you see.

Work together to practice mixing solids and solids. Pour the sprinkles into the dish of ice cream. Mix it with a spoon. **Write** or **draw** what you see.

2.2 | Learn — What changes to objects can be reversed, and what changes cannot?

Record your observations of your mixtures in the charts.

Observations		
Liquid and liquid	Liquid and solid	Solid and solid

Think About the Activity

How were the mixtures the same? How were they different?

Same	Different

What patterns did you notice when you mixed solids with solids, liquids with liquids, and solids with liquids?

_____

_____

_____

2.2: Changing Materials

Activity 9
Analyze Like a Scientist

Quick Code: us2296s

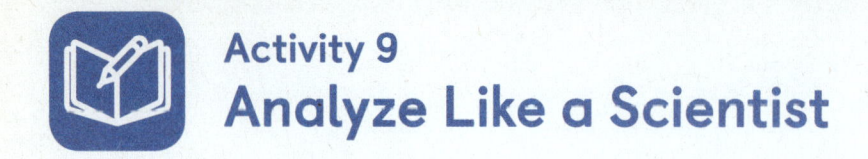

Make a Mixture

Read about mixtures.
Circle examples of mixtures. Then, **answer** the question.

> Read Together

Make a Mixture

We make **mixtures** everyday at home and school. A mixture is a combination of different things. You have probably made hundreds of mixtures in your life.

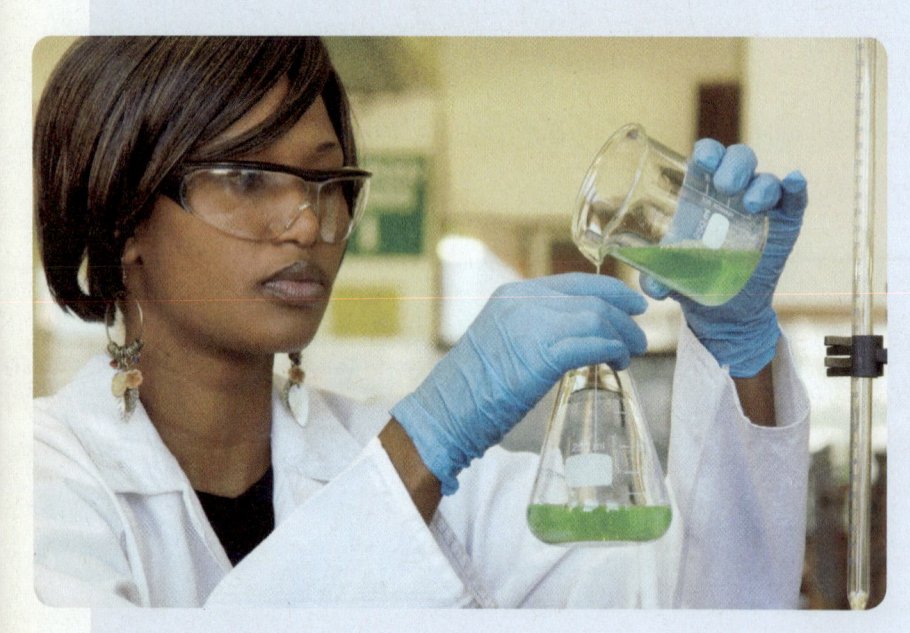

Think about a salad. It is a mixture. A salad with different ingredients is a mixture of solids.

Salads are not the only things that are mixtures of solids. Some rocks are made of pieces of many other rocks. Those rocks are also mixtures.

Mixtures can also be made from liquids. If you mix orange juice and apple juice, you've made a mixture.

The last type of mixture is made from a solid and a liquid. Powdered hot chocolate is a solid. When you stir it into water, you create a mixture of powder and water.

What other mixtures have you made?
_____
_____

2.2: Changing Materials | 91

2.2 | Learn — What changes to objects can be reversed, and what changes cannot?

Activity 10
Observe Like a Scientist

Quick Code: us2297s

Solutions

Complete the activity to learn about **solutions**. **Write** your observations in the chart. Then, answer the questions.

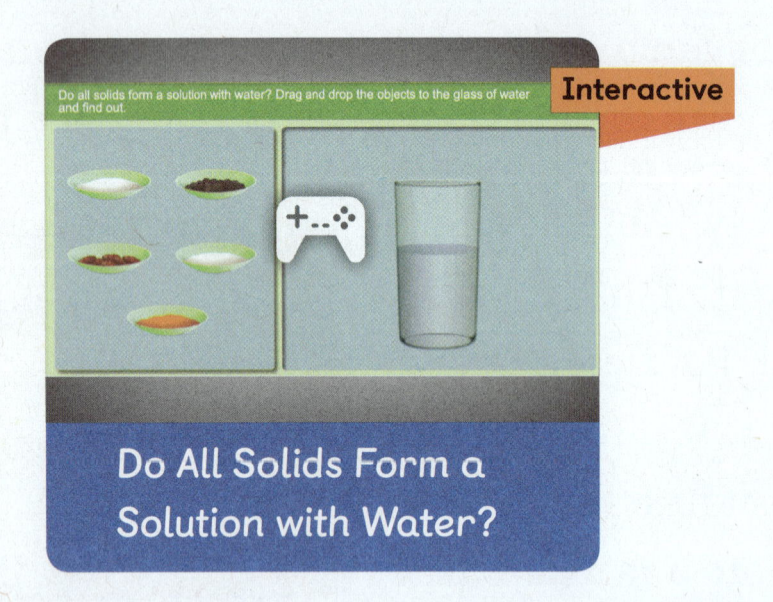

Do All Solids Form a Solution with Water? (Interactive)

Substance	Forms a Solution with Water (Yes/No)
Salt	
Sugar	
Plastic button	
Pebble	
Food coloring	

What happens when you add food coloring to water?

_____
_____
_____
_____

Why does water taste sweeter if you add sugar to it?

_____
_____
_____
_____

How do the physical properties of salt change when you add water to it?

_____
_____

Why might someone want to separate a salt mixture?

_____
_____

SEP Planning and Carrying Out Investigations

2.2: Changing Materials | 93

2.2 | Learn — What changes to objects can be reversed, and what changes cannot?

Activity 11
Observe Like a Scientist

Quick Code: us2298s

Building Materials

Watch the video. **Look** for ways that mixtures are made. Then, **talk** about what you observed.

Building Materials (Video)

Talk with a partner about what you learned from the video. As you share each statement, place an X or an O to show you have shared your thinking.

After your partner has shared his or her thinking, **share** a piece of evidence from the video that connects to his or her thinking.

XO Let's Go

2.2: Changing Materials

2.2 | Learn — What changes to objects can be reversed, and what changes cannot?

Activity 12
Observe Like a Scientist

Quick Code: us2299s

Rocks Are Mixtures

Watch the video. Pay attention to why a rock is a mixture. Then, **talk** about what you observed.

Video

Rocks Are Mixtures

Talk with a partner about what you learned from the video. As you share each statement, place an X or an O to show you have shared your thinking.

After your partner has shared his or her thinking, **share** a piece of evidence from the video that connects to his or her thinking.

XO Let's Go

2.2: Changing Materials

Activity 13
Analyze Like a Scientist

Quick Code: us2300s

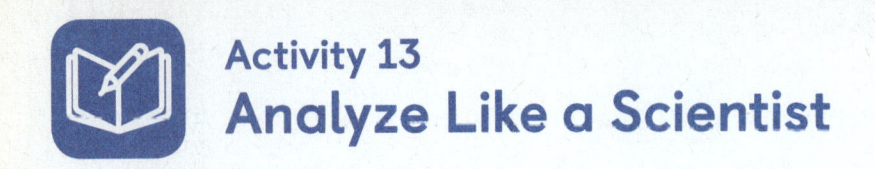

Mixing Materials

Read the text. **Look** for vocabulary words. Then **answer** the questions that follow.

Read Together

Mixing Materials

Sometimes, we can put objects together without changing them. The colorful river pebbles have been combined to form a mixture. A mixture is a combination of things. Each part can be separated from the other things. From the river pebbles mixture, you can pick out all the blue pebbles. The blue pebbles are still the same. The rest of the pebble mixture is still the same. What other objects can you mix together?

River Pebbles

Sometimes when you mix together different objects, it is easy to see each object in the mixture. But what happens when you mix water, lemons, and sugar to make a drink like lemonade? When you stir in the sugar, it seems to disappear. The sugar will **dissolve** in the water. You know the sugar is still there because you can taste it. A solid dissolved in a liquid is called a solution.

Lemonade

2.2: Changing Materials

Read Together

There are many mixtures we can eat and drink. A salad can be a mixture of different vegetables. Lemonade is a mixture of water, lemon juice, and sugar. But did you know that there are also many useful mixtures that can be made using landscape **materials** like rocks and **soil**? Concrete and bricks are two examples.

Pouring Concrete

What is a mixture?
_____
_____

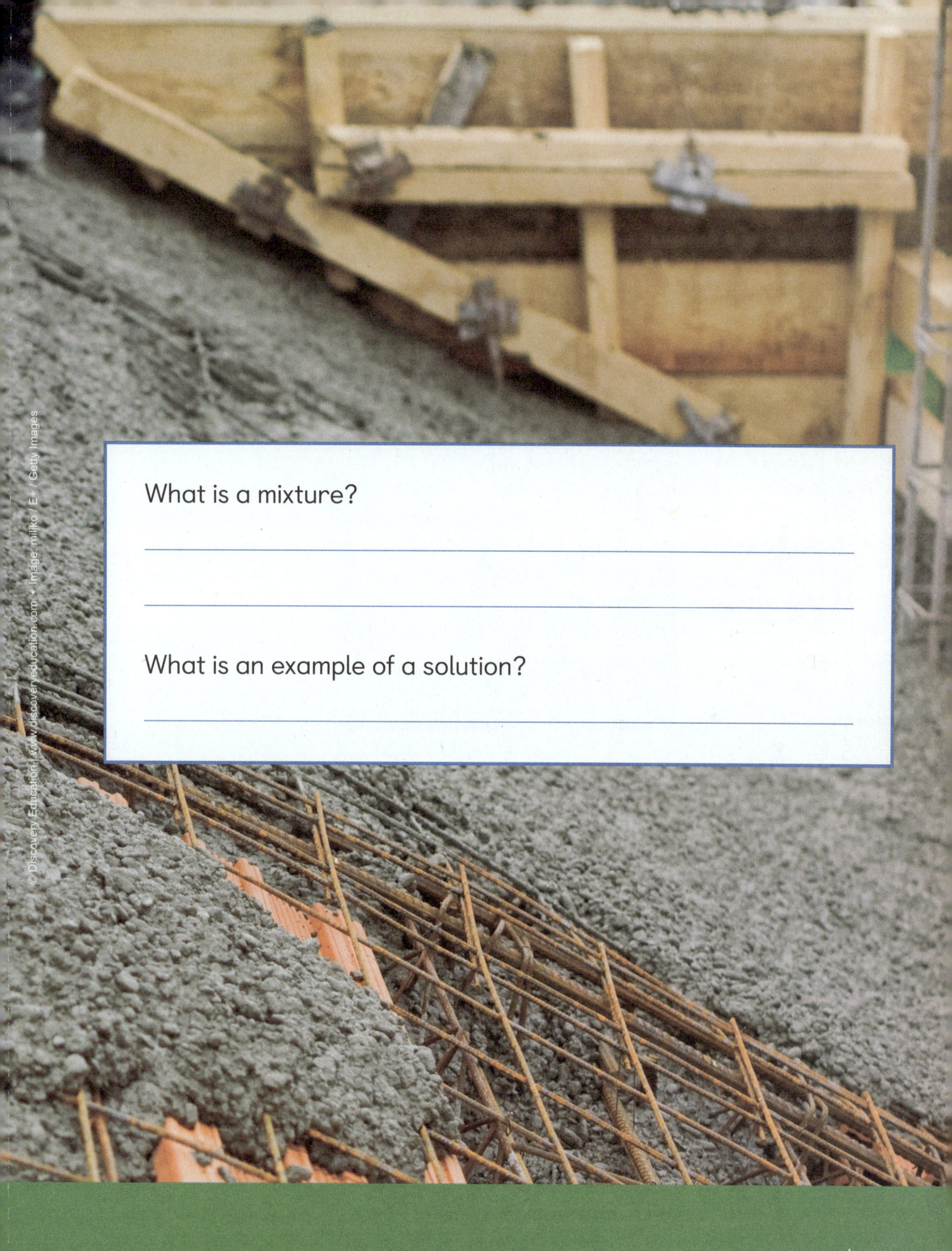

What is an example of a solution?
_____

2.2: Changing Materials | 101

2.2 | Learn   What changes to objects can be reversed, and what changes cannot?

Activity 14
Observe Like a Scientist

Quick Code: us2301s

Music Video: Mixtures

Watch the video. **Look** for examples of mixtures. Then, **talk** about what you observed.

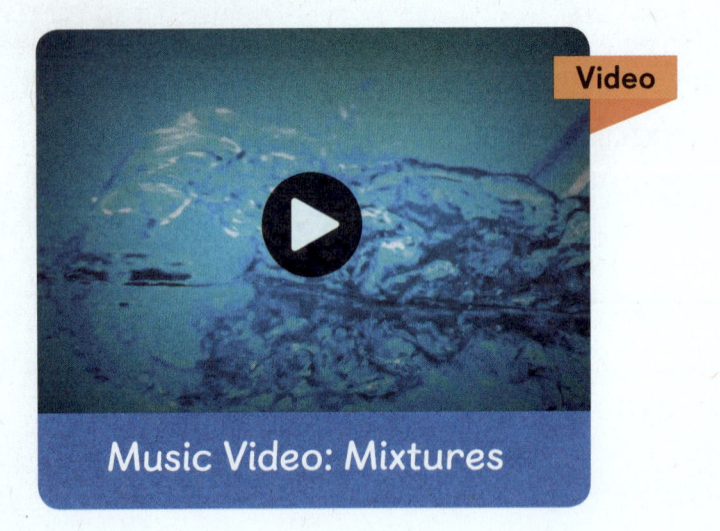

Video

Music Video: Mixtures

Talk Together

Now, talk together about motions or dance moves that go with the music in the video.

CCC Stability and Change

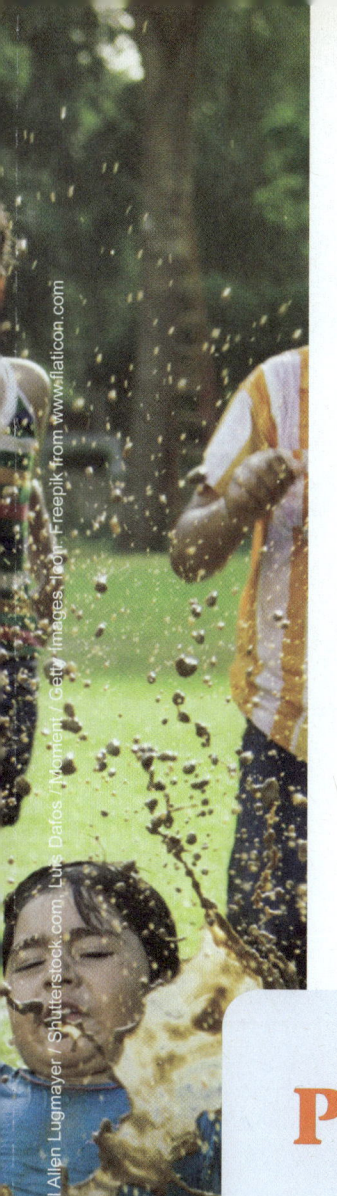

How Can Materials in a Mixture Be Pulled Apart Again?

Activity 15
Analyze Like a Scientist

Quick Code: us2302s

Playing in the Mud

Read the text. **Look** at the photo. Then, **answer** the questions.

Read Together

Playing in the Mud

It is fun to play in mud. Mud is a mixture of soil and water. But suppose you want to play another game. It is hard to play baseball in the mud. Is it possible to separate the soil from the water so the ground is dry again?

Mud Puddle

2.2: Changing Materials | 103

Is it possible to separate the soil from the water so the ground is dry again?

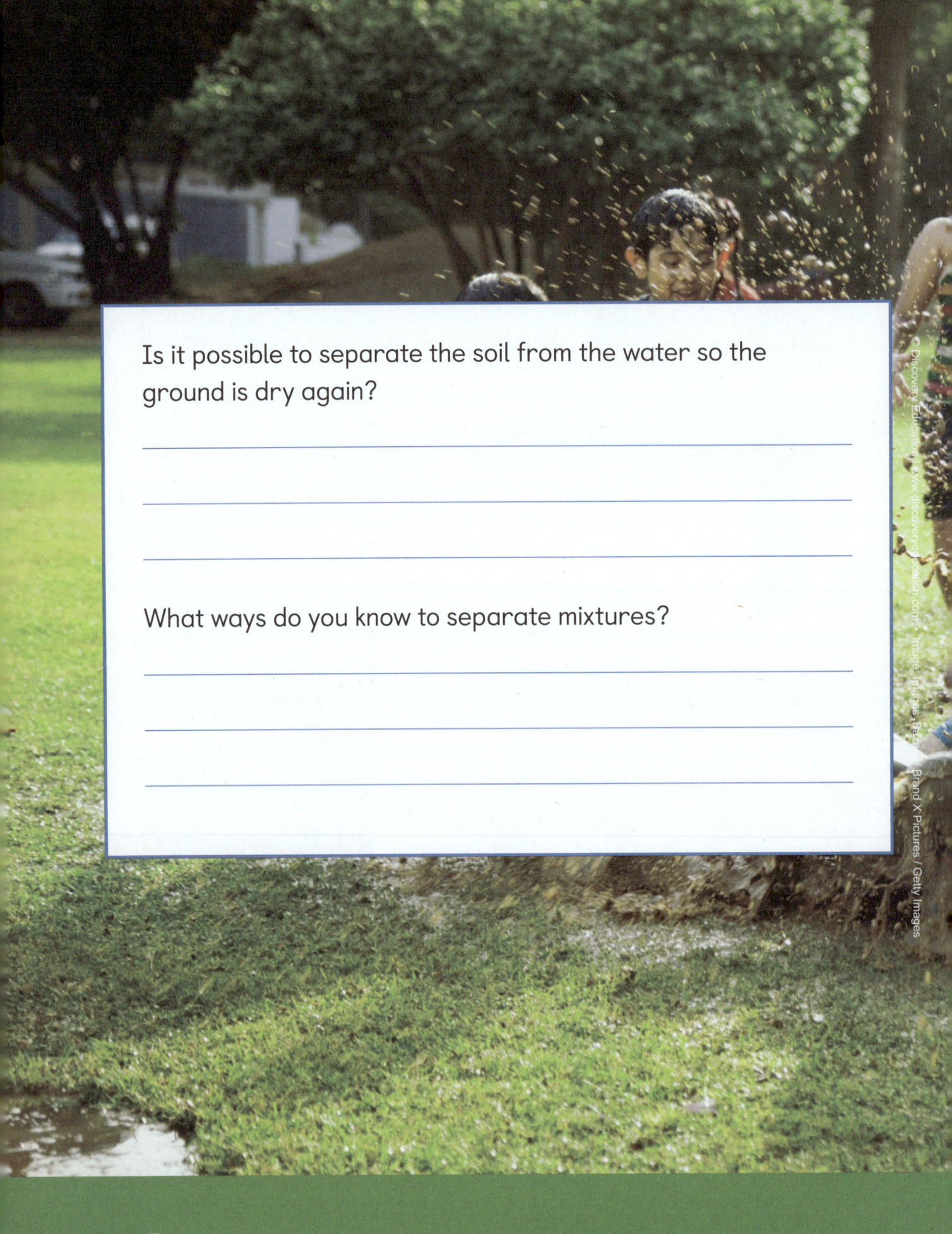

_____
_____
_____

What ways do you know to separate mixtures?

_____
_____
_____

Activity 16
Investigate Like a Scientist

Quick Code: us2303s

Hands-On Investigation: Separating a Mixture

In this activity, you will take apart, or separate, a mixture. A mixture is two or more things put together. You can separate a mixture into its parts.

Make a Prediction

You are going to separate a mixture of sand and rice or beans and rice.

Write or **draw** your predictions.

Do you think it will be easy to separate the mixture? Why or why not?

SEP Planning and Carrying Out Investigations **CCC** Patterns

2.2: Changing Materials

2.2 | Learn What changes to objects can be reversed, and what changes cannot?

What materials do you need? (per group)

- Seeds, kidney beans
- Plastic cup, 5 oz
- Netting (piece), 2 × 2 feet
- Rice
- Spoon, plastic
- Strainers
- Sand
- Forceps

HANDS-ON INVESTIGATION

What do you think the mixture will look like after it has been separated?

What Will You Do?

Think about how you will separate the mixture. What tools will you use? **Draw** a picture.

2.2: Changing Materials | 107

2.2 | Learn — What changes to objects can be reversed, and what changes cannot?

Think About the Activity

How did you separate your mixture? How did the mixture change? **Draw** a picture.

Before Separating	After Separating

Did the parts of the mixture look the same or different after you separated them? Explain.

_____

_____

_____

What are some mixtures you can find at home or in nature? **Write** or **draw** pictures.

2.2: Changing Materials | 109

2.2 | Learn What changes to objects can be reversed, and what changes cannot?

Activity 17
Observe Like a Scientist

Quick Code: us2304s

Recycling and Sorting

Watch the video. **Look** for differences in the materials that are mixed together. Then, **talk** about what you observed.

Recycling and Sorting

Talk Together

Now, talk together about which tool would be the best to use to separate the paper clips from the mixture.

SEP Engaging in Argument from Evidence

Activity 18
Observe Like a Scientist

Quick Code: us2305s

Mixtures: Paul and His Science Pals

Watch the video. **Look for** how the mixture of soil and water are separated. Then, **talk** about what you observed.

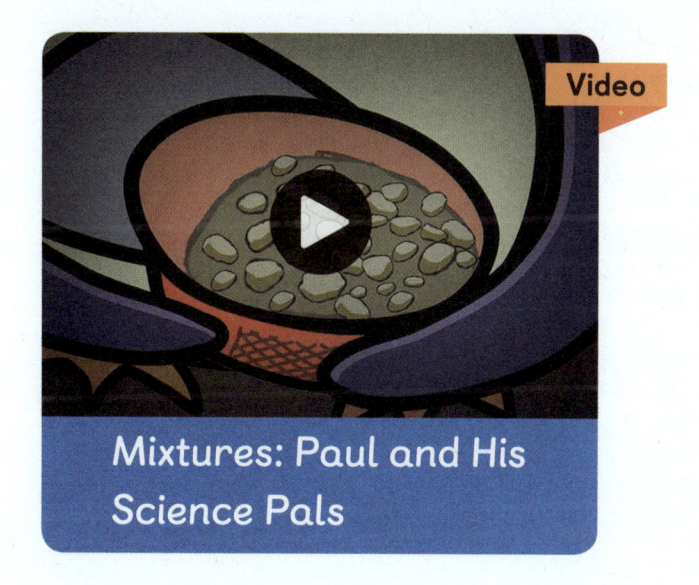

Mixtures: Paul and His Science Pals

Talk Together

Now, talk together about what evaporation is and what causes it.

SEP Engaging in Argument from Evidence

2.2: Changing Materials | 111

2.2 | Learn — What changes to objects can be reversed, and what changes cannot?

Activity 19
Investigate Like a Scientist

Quick Code: us2306s

Hands-On Investigation: Looking at Layers

In this activity, you will make a model. It will show you how layers settle out of water on Earth's surface.

Make a Prediction

You are going to put soil, leaves, water, and sand in a bottle. You will seal the bottle and shake it.
Write or **draw** your predictions.

What do you think will happen to the different materials?

SEP Planning and Carrying Out Investigations
CCC Patterns

What materials do you need? (per group)

- Plastic bottle, $1\frac{1}{2}$ L
- Crayons
- Funnel
- Leaves
- Gravel
- Water
- Pencils
- Sand
- Soil, potting

HANDS-ON INVESTIGATION

What Will You Do?

Put the materials in the bottle. Use the funnel if needed. Put the cap on the bottle. Shake the bottle. What happened?

_____

_____

_____

_____

2.2: Changing Materials

2.2 | Learn — What changes to objects can be reversed, and what changes cannot?

Think About the Activity

How did the mixture change when you shook the bottle? **Write** or **draw** to show what happened.

Before	After

Changes

Activity 20
Observe Like a Scientist

Quick Code: us2307s

Mixtures

Complete the activity to make and separate mixtures. **Write** your observations in the chart. Then, **answer** the questions.

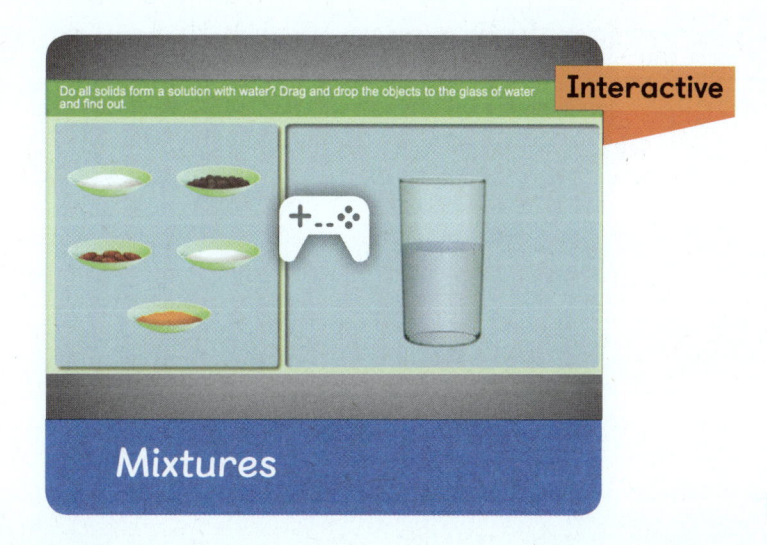

Interactive

Mixtures

Mixture	Tools Used to Separate the Mixture	Property Used to Separate the Mixture
Sand and iron filings		
Oil and sand		
Water and salt		

2.2: Changing Materials | 115

2.2 | Learn — What changes to objects can be reversed, and what changes cannot?

Why won't a magnet let you separate a mixture of salt and water?

_____
_____
_____

Give two examples of mixtures in which the physical properties of the materials DO NOT change when they are mixed together. Then give two examples of mixtures in which the physical properties of the materials DO change when they are mixed together.

_____
_____
_____
_____
_____

Can you separate a mixture of sand and salt with a filter? Why or why not?

_____
_____

Think about the difference between oil and water. Think about what happens when you mix oil and water. How could you separate a mixture of oil and water?

_____
_____

Activity 21
Observe Like a Scientist

Quick Code: us2308s

A Dry Lake in the Mojave Desert

Look at the photo.

A Dry Lake in the Mojave Desert

Draw a picture of what the lake would have looked like before the water evaporated.

2.2: Changing Materials

Activity 22
Analyze Like a Scientist

Quick Code: us2309s

How Can Mixtures Be Separated?

Read the text. **Underline** ways to separate mixtures.

> **Read Together**
>
> ## How Can Mixtures Be Separated?
>
> Objects in a mixture can be separated. Individual parts of a mixture can be separated from the other parts. What are some strategies you might use to separate the materials in a mixture?
>
> Mixtures can be separated in a variety of ways. Understanding the properties of the materials in a mixture can help you figure out which separation method would work best.

SEP Engaging in Argument from Evidence

Use a Filter

Use a Magnet

These are a few strategies for separating mixtures:

- Use a filter. Small objects will fit through a filter. Bigger objects will not.
- Add water to a mixture. Some objects may float, while others may sink.
- Add heat to a water mixture. The water will evaporate, while solid particles are left behind.
- Use a magnet. Some objects are attracted to magnets, while others are not.

2.2: Changing Materials | 119

Write or **draw** about a time that you separated a mixture.

Activity 23
Evaluate Like a Scientist

Quick Code: us2310s

Mixed or Separated?

Some of the objects have been mixed, while others have been separated. Write **Mixed** or **Separated** below each picture.

_____ _____

_____ _____

SEP Engaging in Argument from Evidence

2.2: Changing Materials | 121

2.2 | **Learn** What changes to objects can be reversed, and what changes cannot?

How Does Temperature Change the Things That Make Up a Material?

Activity 24
Observe Like a Scientist

Quick Code: us2311s

Adobe House

Look at the photo to find materials the adobe house is made of.

Adobe House

Talk Together

Now, talk together about why this type of home would be found in places where it is hot.

122

Activity 25
Analyze Like a Scientist

Quick Code: us2312s

Adobe Bricks

Read the text. Then, answer the question.

Read Together

Adobe Bricks

Bricks are a useful material for building houses. To make adobe bricks, mix together clay, sand, water, straw, and one final ingredient. What else needs to be added to turn this soft mixture into a hard brick?

Adobe Bricks

2.2: Changing Materials | 123

Read Together

Heat from the sun changes the properties of an adobe mixture. The heat turns the soft, wet mixture into hard, dry brick.

Dry Bricks

Think back to the three little pigs. What could the little pig add to the straw to make his house stronger?

_____
_____

Activity 26
Observe Like a Scientist

How to Make Adobe Bricks

Watch the video. **Look** for ways that soil is used to make houses. Then, **talk** about what you observed.

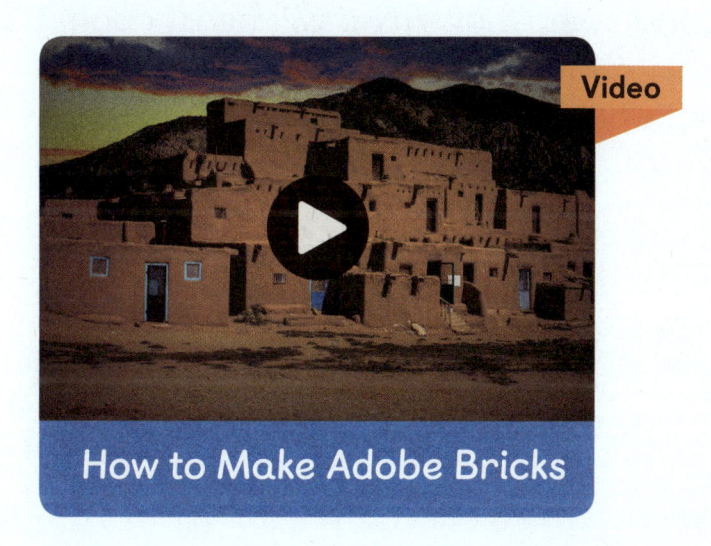

Quick Code: us2313s

How to Make Adobe Bricks

Talk Together

Now, talk together about what is in the mixture that makes adobe bricks, and what makes the bricks become hard.

2.2: Changing Materials

2.2 | Learn — What changes to objects can be reversed, and what changes cannot?

Activity 27
Investigate Like a Scientist

Quick Code: us2314s

Hands-On Investigation: Slow the Flow

In this activity, you will observe how temperature changes the flow of syrup. One sample of syrup will be frozen and then thawed. A second sample of syrup will be at room temperature.

Make a Prediction

You are going to observe the flow of syrup. Each syrup sample will be a different temperature.

Do you think the temperature of the syrup will change how it flows? Why or why not?

Write or **draw** your predictions.

SEP Planning and Carrying Out Investigations

What materials do you need? (per group)

- Light corn syrup
- Markers
- Measuring cups
- Plastic cup, 9 oz
- A way to freeze 1 cup of syrup
- Pencils
- Plastic container, 12 oz
- Thermometer, plastic
- Disposable tablecloth

HANDS-ON INVESTIGATION

What Will You Do?

How will you do your investigation? How will you make the syrup samples two different temperatures? What will you observe? Write or draw to show what you will do.

2.2: Changing Materials

2.2 | Learn What changes to objects can be reversed, and what changes cannot?

You will pour syrup from a container and watch it flow. Is syrup a solid, liquid, or gas?

_____

What state of matter is syrup if you freeze it?

_____

Pour each syrup sample. Write or draw to tell what happens.

Think About the Activity

Write or **draw** the answers to the questions in the chart. Did your two syrup samples pour the same? Explain. What patterns did you notice?

Did Samples Pour the Same?	Explain

What Patterns Did You Notice?

2.2: Changing Materials | 129

2.2 | Learn — What changes to objects can be reversed, and what changes cannot?

How Can We Describe and Organize the Changes That Happen to a Material?

Activity 28
Investigate Like a Scientist

Quick Code: us2315s

Hands-On Investigation: Changing Clay

In this activity, you will change the shape of clay by making an object of your choice.

Make a Prediction

You are going to change the shape of clay using cutting and shaping tools.

What properties of the clay do you think will change?

Write or **draw** your predictions.

SEP Planning and Carrying Out Investigations

What materials do you need? (per group)
- Digital camera
- Disposable tablecloth
- Modeling clay
- Refrigerator
- Knife, plastic

What Will You Do?

Use the tools and your hands to change the shape of your clay. **Make** an object of your choice.

How does the clay change as you work with it?

_____

_____

_____

_____

2.2 | Learn — What changes to objects can be reversed, and what changes cannot?

Think About the Activity

Write or **draw** to show how your clay changed.

Before	After

Activity 29
Observe Like a Scientist

Quick Code: us2316s

Dried Clay

Watch the video. **Look** for how air drying changes clay. Then, **talk** about what you observe.

Video

Dried Clay

Talk Together

Now, talk together about what would happen if the clay is heated in a kiln. Can the change be reversed?

2.2: Changing Materials | 133

2.2 | Learn What changes to objects can be reversed, and what changes cannot?

Activity 30
Observe Like a Scientist

Quick Code: us2317s

Fired Clay

Watch the video. **Look** for evidence that heated clay is reversible or irreversible. Then, **talk** about what you observe.

Fired Clay (Video)

Talk Together

Now, talk together about the story of the Three Little Pigs. Talk about why the third pig's house was so strong.

Activity 31
Analyze Like a Scientist

Quick Code: us2318s

Reversible and Irreversible Changes

Read the statements. Three are true, and one is a lie. After you **read** the passage, **circle** the statement that is false.

- A hard clay pot that is air-dried can be changed back into a lump of clay.

- When a material can be changed back to its original form, this is called reversible change.

- A clay pot that has been heated in an oven is an irreversible change.

- Burnt toast is an example of a reversible change.

2.2: Changing Materials

Read Together

Reversible and Irreversible Changes

Reversible Change

A clay pot that is air-dried can be changed back into a lump of clay. Some material can be changed back to its original form. This is called a reversible change. Water can change from solid to liquid and back to solid again. This is an example of a reversible change.

Once a clay pot has been heated or fired in an oven, can the clay be changed back to its original form?

Soft clay can be fired in an oven. When it comes out of the oven, it is not soft anymore. It is impossible to change the clay back to its original form. Some material cannot be changed back to its original form. This is called irreversible change. A toaster turning soft bread into crispy bread is an example of an irreversible change.

Irreversible Change

2.2: Changing Materials

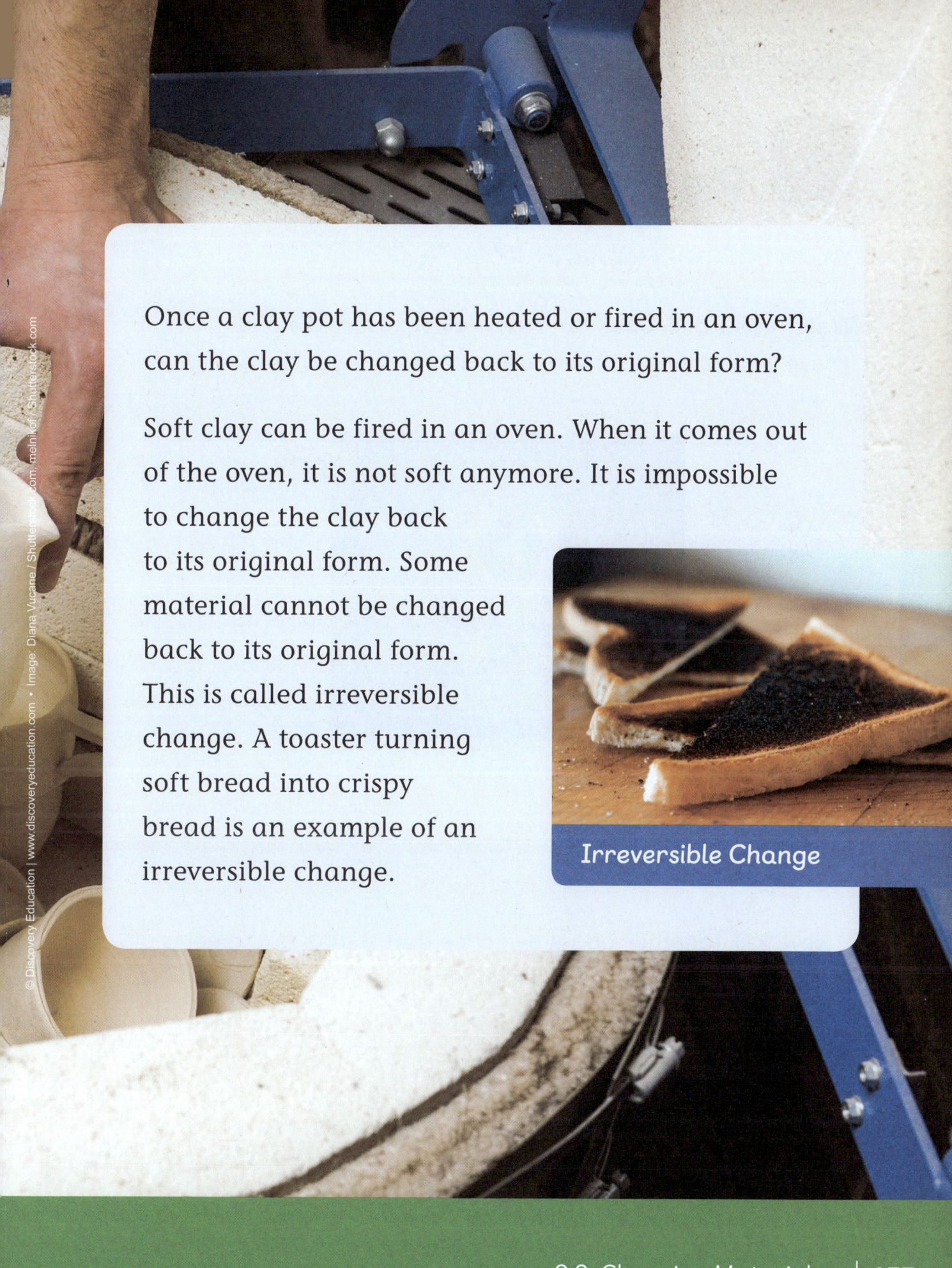

2.2 | Learn — What changes to objects can be reversed, and what changes cannot?

Activity 32
Observe Like a Scientist

Quick Code: us2319s

Freezing and Melting

Watch the video. **Look** for examples of reversible and irreversible changes. Then, **talk** about what you observed.

Video

Freezing and Melting

Talk Together

Now, talk together about evidence that supports examples of reversible and irreversible changes.

CCC Stability and Change

Activity 33
Evaluate Like a Scientist

Quick Code: us2320s

How Has It Changed?

Study each set of pictures. They each show a change to matter. In the paragraph, **write** the word that best completes the sentence about the changes.

waterfall	apple	egg

The _____ has changed from whole to parts. This is a change in shape. The _____ has changed from uncooked to cooked. This is a change in texture and temperature. The _____ has changed from liquid to solid. This is a change in state.

SEP Engaging in Argument from Evidence

2.2: Changing Materials

2.2 | Share
What changes to objects can be reversed, and what changes cannot?

Activity 34
Record Evidence Like a Scientist

Quick Code: us2321s

A Smart Little Pig

Now that you have learned about changing materials, look again at A Smart Little Pig. You first saw this in Wonder.

Let's Investigate A Smart Little Pig

Talk Together

How can you describe A Smart Little Pig now?
How is your explanation different from before?

SEP Constructing Explanations and Designing Solutions

Look at the Can You Explain? question. You first read this question at the beginning of the lesson.

> ### Can You Explain?
>
> What changes to objects can be reversed and what changes cannot?

Now, you will use your new ideas about A Smart Little Pig to answer a question.

1. **Choose** a question. You can use the Can You Explain? question, or one of your own. You can also use one of the questions that you wrote at the beginning of the lesson.

My Question

2. Then, use the sentence starters on the next page to **answer** the question.

2.2: Changing Materials | 141

2.2 | Share — What changes to objects can be reversed, and what changes cannot?

I know that material changes can be different because
_____
_____
_____
_____

Based on my observations,
_____
_____

can change materials because
_____
_____
_____

An example of a reversible change would be

_____

_____

_____

_____

An example of an irreversible change would be

_____

_____

_____

_____

The evidence I collected shows

_____

_____

_____

_____

2.2: Changing Materials | 143

STEM in Action

Quick Code: us2322s

Activity 35
Analyze Like a Scientist

Changing Clay for a Job

Read about potters.

Read Together

Changing Clay for a Job

Can you imagine a job in which you get paid for changing clay into pots? Potters take a big block of clay and make it into a pot. First they have to make the block smaller. Then they have to make the clay thinner or into a ball shape.

A potter changes clay. The potter makes a shape out of clay. Then the clay is heated. The clay changes from soft to hard.

Making Pottery

Think about what you have read. Then, **watch** the video and **answer** the questions.

Video

How Can Matter Change?

How does heat change clay?

_____

_____

What tools does a potter use?

_____

_____

SEP Obtaining, Evaluating, and Communicating Information

CCC Structure and Function

2.2: Changing Materials | 145

Make Clay Smaller

You are a potter. You get a big block of clay. You must show someone how to make the block into smaller pieces of different shapes. **Draw** a line to match each tool with the shape it can make. Some tools can make more than one shape.

roller	ball
hands	flat sheet
knife	strips
	long rope
	little squares

Activity 36

Evaluate Like a Scientist

Quick Code: us2323s

Review: Changing Materials

Think about what you have read and seen.
What did you learn?
Draw what you have learned.
Then, **tell** someone else about what you learned.

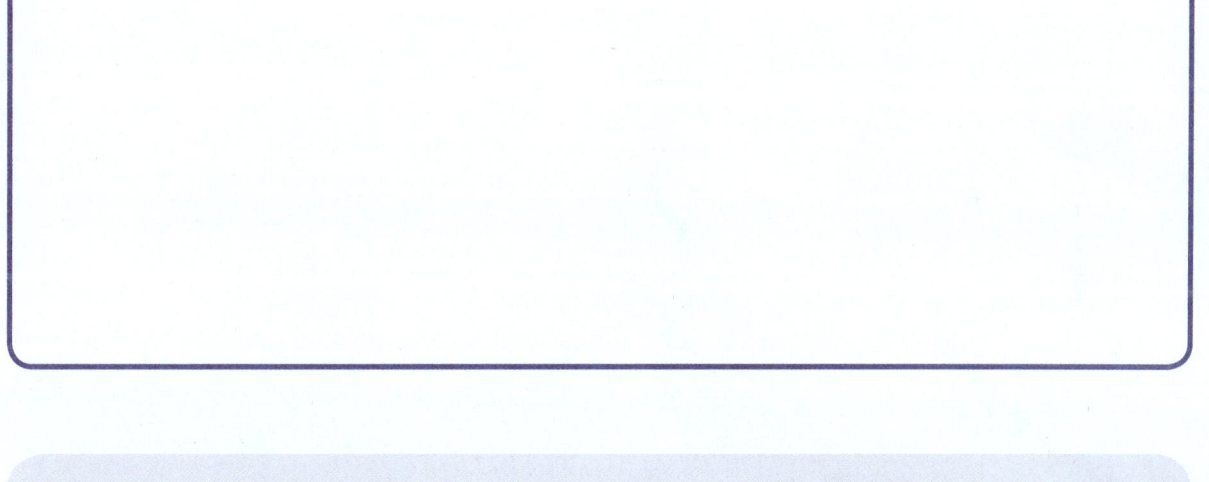

Talk Together

Think about what you saw in Get Started. Use your new ideas to discuss changes in materials.

SEP Obtaining, Evaluating, and Communicating Information

2.2: Changing Materials | 147

CONCEPT
2.3

Materials in Design

Student Objectives

By the end of this lesson:

- ☐ I can explore and test materials to find which material properties are best for a purpose.

- ☐ I can take apart an object and design and build a new object out of the parts.

- ☐ I can explore data tests to compare the strengths and weaknesses of a given design solution.

Key Vocabulary

- ☐ absorption
- ☐ gemstone
- ☐ landfill
- ☐ matter
- ☐ resource
- ☐ reverse engineering

Quick Code: us2325s

Activity 1
Can You Explain?

How are materials chosen for designs?

_____

_____

_____

_____

_____

Quick Code:
us2326s

2.3: Materials in Design | 151

2.3 | Wonder — How are materials chosen for designs?

Activity 2
Ask Questions Like a Scientist

Quick Code: us2327s

Three Pigs Problem

Watch the video. **Look** for the properties of the materials that the pigs used.

Let's Investigate Three Pigs Problem

SEP Planning and Carrying Out Investigations

Which material had the best properties for building? Why or why not?

_____

_____

_____

Log Cabin Model

Talk Together

Now, talk together about the properties of the materials used for the Little Pigs' house. Talk about the properties of the materials used for the model house. Talk about what questions you want to ask about the materials.

2.3: Materials in Design | 153

Activity 3
Analyze Like a Scientist

Quick Code: us2328s

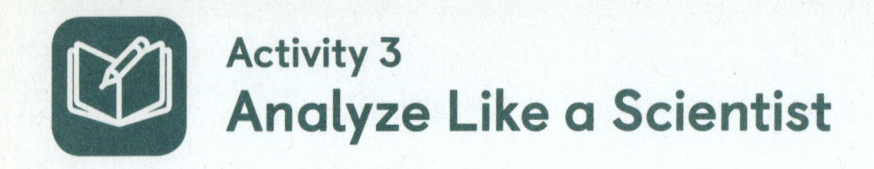

Materials in Design

Read the text. Then complete the chart.

Read Together

In the story of the Three Little Pigs, the first pig finds straw in his local landscape. He uses the straw to build a house. After the wolf blows his house down, the pig needs to stop and think. Was straw the best material for the job?

The first little pig could look at a model of a wooden log house. He could take apart the model to see how it was put together. Then, the pig could build a similar house. Is it possible to use a different material to re-create the design of the wooden log house? What other materials might the first little pig use to build a house?

Log Cabin Model

154

Here is an idea for the first little pig. Instead of using straw that is dry grass, maybe he could use a different type of straw to build a house. Is it possible to make a house using drinking straws? You can make a model to find out!

Drinking Straws

What kinds of different materials can be used to make a model house?

I see...	I think...	I wonder...

Activity 4
Design Solutions Like a Scientist

Quick Code: us2329s

Hands-On Engineering: Rebuilding a House

In this activity, you will use the concept of **reverse engineering** to make a house. You will take apart a house of wooden blocks and try to re-create the design using cardboard and straws.

Ask Questions About the Problem

You are going to observe a house made of wooden blocks before you take it apart and re-create the design with cardboard and straws.

What questions do you have about the problem? **Write** or **draw** your questions.

SEP Planning and Carrying Out Investigations
CCC Patterns

2.3 | Wonder How are materials chosen for designs?

What materials do you need? (per group)

- Pre-constructed house made of wooden blocks
- Cardboard
- Plastic straws
- Glue
- Black permanent marker
- Metric ruler
- Scissors
- Pencils
- Large straws
- Straws made of different materials (paper, plastic, etc.)

HANDS-ON ENGINEERING

What Will You Do?

Observe the house made of wooden blocks. **Write** or **draw** what you see.

Take the house apart to put it back together. Then **talk** in your group about how you want to build your house using the available materials. **Talk** about what parts of the house you need to build. Talk about any special features that you need to add. **Talk** about how to be sure the pieces stay together and the house stays standing.

Think About the Activity

How did you build your house? **Write** and **draw** a picture.

How I Used the Materials	Picture of House

2.3: Materials in Design

Activity 5
Analyze Like a Scientist

Quick Code: us2330s

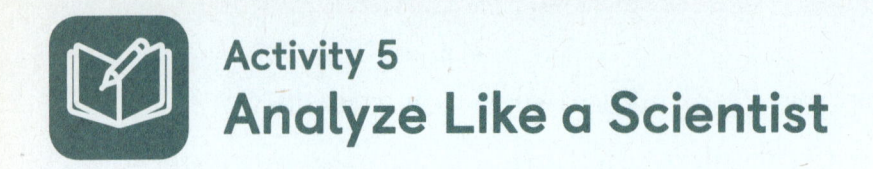

Materials and Systems in a Home

Read the text. **Look** for the materials that are used when building a house. **Circle** the materials used.

Read Together

Materials and Systems in a Home

Kim's dad is building a new house. She is keeping a journal.

Concrete Blocks and Cement

August 12

Today, Dad made a base for our house. He used concrete blocks. Cement holds them together.

August 20

Dad is building a frame for the house. The wood is held together with metal nails. They will make the walls strong.

August 30

The walls are up!
Dad put in glass windows.
The outside is covered in stones.

Wood frames help make the walls of a house. They are part of the structural system of a house.

September 5

People are working in the house.
An electrician uses copper wires.
This will make our lights work.
A plumber uses plastic pipes.
These pipes will bring us water.

October 1

Our house is finally finished.
Dad used many different materials!

Wires are part of the electrical system and pipes are part of the plumbing system in a home.

2.3: Materials in Design

2.3 | Wonder How are materials chosen for designs?

Activity 6
Evaluate Like a Scientist

Quick Code: us2331s

What Do You Already Know About Materials in Design?

Materials in My Kitchen

Think about the different things that you have in your kitchen. There might be a toaster, napkins, and a refrigerator. **Pick** three different things and **name** at least one material that they are made of. Why do you think they have this material?

What Material Is It?

Look at each object. What material is used most in each one? **Draw** a line to match each object to the correct material.

Object	Material
	plastic
	wood
	metal

2.3: Materials in Design | 163

How Do Engineers Choose Materials?

Activity 7
Analyze Like a Scientist

Quick Code: us2332s

Choosing Materials

Read the text. Then, **answer** the question.

> Read Together

Choosing Materials

If you look around your house, you will see objects made of many different materials. Some objects may be made of plants, like trees. Other objects may be made of metal. Some objects may be made of rocks.

Kitchen Objects

Objects are designed for a purpose. An object's material must help it achieve this purpose. Paper mittens would absorb too much water. They would not protect your hands from snow. A metal jacket would not be flexible enough to wear.

Engineers choose materials with properties that will solve problems. Suppose an area does not get a lot of rain. The farmers want a soil that will hold enough water for plants to grow. Engineers will choose materials that absorb the necessary water when designing a new soil mixture.

What problem do engineers need to solve for farmers who live in areas that do not get a lot of rain?

_____

_____

_____

_____

_____

2.3: Materials in Design | 165

2.3 | Learn — How are materials chosen for designs?

Activity 8
Investigate Like a Scientist

Quick Code: us2333s

Hands-On Investigation: Experimenting with Soil

In this activity, you will test different types of soil to see how well they hold water. You will be able to plan and conduct the investigation yourself.

Make a Prediction

You are going test three different kinds of soil to see how well they hold water. **Write** or **draw** your predictions.

What kind of soil do you think will hold the most water?

What kind of soil do you think will hold the least water?

What materials do you need? (per group)

- Soil, potting
- Plastic bottle, 8 oz
- Graduated cylinder, 50 × 1 mL
- Beakers, plastic, 250 mL
- Dish soap
- Paper towels
- Disposable gloves (per student)
- Balance, double pan
- Plastic cup, 9 oz
- Calculator, battery-powered
- Plastic pan

What Will You Do?

How will you test the soils? What information will you collect? **Write** or **draw** what you will do.

SEP Planning and Carrying Out Investigations
CCC Patterns

2.3: Materials in Design | 167

2.3 | Learn How are materials chosen for designs?

Write your data in the chart.

	Soil 1	Soil 2	Soil 3	Soil 4
A. Mass of soil (g) *before* pouring water				
B. Mass of soil (g) *after* pouring water				
Amount of water absorbed (mL): column B – column A				
Amount of water remaining (mL)				

Think About the Activity

Which type of soil holds water the most water? Which type holds the least water? Show what happened. **Write** and **draw** your answers in the chart.

Held Most Water	Held Least Water

2.3: Materials in Design | 169

How Can Materials Be Tested?

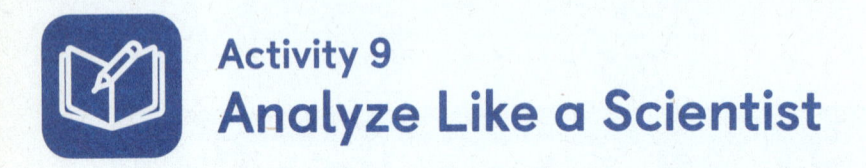

Activity 9
Analyze Like a Scientist

Quick Code: us2334s

Testing Materials

Read the text. **Underline** the steps an engineer would take to test materials.

> **Read Together**
>
> # Testing Materials
>
> Engineers compare the properties of materials by testing them. Here are three basic steps for testing materials:
>
> 1. Decide the property you want to test. For example, you could test for the flexibility, **absorption**, or strength of an object.
> 2. Choose two or more different materials or objects to test.
> 3. Test the materials. Tests should be fair. Test each material or object the same way.

2.3: Materials in Design

2.3 | Learn — How are materials chosen for designs?

Activity 10
Observe Like a Scientist

Quick Code: us2335s

Blow the House Down: Stucco in Straw As an Efficient Building Material

Watch the video. **Fill in** the chart that follows.

Blow the House Down: Stucco in Straw As an Efficient Building Material

SEP Planning and Carrying Out Investigations

Building Materials Investigation

	Fireproof	Waterproof	Thermal (Heat) Resistant
Cinder Block			
Straw Bale			
Straw and Stucco			

Talk Together

Now, talk together about whether any of the results of the building materials investigation were surprising.

2.3: Materials in Design

2.3 | Learn — How are materials chosen for designs?

Activity 11
Investigate Like a Scientist

Quick Code: us2336s

Hands-On Investigation: The Great Ice Cube Race

In this activity, you will predict which material will keep an ice cube from melting the longest. Then, you will test the materials and compare the results with your predictions.

Make a Prediction

You are going to put ice cubes in cups made of foam, metal, plastic, and paper. **Write** or **draw** your predictions.

Which material do you think will keep the ice cube from melting the longest? Why?

SEP Planning and Carrying Out Investigations
CCC Patterns

What materials do you need? (per group)

- Ice cubes
- Stopwatch
- Foam cup, 14 oz
- Lid for foam cup
- Can, black metal
- Lid for metal cup
- Plastic cup, 9 oz
- Lid for plastic cup
- Waxed paper cup, 200 mL
- Lid for paper cup

HANDS-ON INVESTIGATION

What Will You Do?

Place one ice cube in each cup.

Cover the cup with a lid.

Fill in the Materials and Race Prediction columns of the chart.

2.3: Materials in Design

2.3 | Learn How are materials chosen for designs?

Cup	Material	Race Prediction	
A			
B			
C			
D			

Wait 10 minutes. Observe the ice cubes. Fill in the rest of the chart.

	Race Result	Strengths	Weaknesses

2.3 | Learn How are materials chosen for designs?

Think About the Activity

Which material would work best in a thermos for holding a cold drink? Why? Which material would be the worst in a thermos? Why? **Write** your answers in the chart.

Works best in a thermos:	Because:
Works worst in a thermos:	Because:

Do you think a thermos could keep an ice cube from melting all day long, or maybe for a week or a month? Explain.

_____

_____

_____

What might happen if you replaced the ice cubes with something hot, like hot chocolate? Do you think the results of the activity would be the same?

_____

_____

_____

Does the pattern in the data support the conclusion that foam prevents ice melting? Why or why not?

_____

_____

_____

Activity 12
Analyze Like a Scientist

Quick Code: us2337s

Max's Strange Objects

Read the text. **Think** about what material would be a better choice for Max's objects. **Write** your choice next to each picture.

Read Together

Max's Strange Objects

Today Max explored the attic. He found an old box and opened it. In the box he found very strange objects. Here are the strange objects Max found.

Max found some red mittens. They were made from paper. The mittens did not keep Max's hands warm.

Max found some crayons. The crayons were made of glass. Max could not draw with them.

Max found a jacket. It was made of metal. Max could not move in the jacket.

Max found some sunglasses. The glasses were made of cotton. Max could not see through them.

What is strange about each of the objects that Max found?

2.3 | Learn   How are materials chosen for designs?

How Can Taking Things Apart Help You When Designing?

Activity 13
Observe Like a Scientist

Quick Code: us2338s

Gemstones

Look at the picture of the **gemstones**. Then, **answer** the question.

Gemstones

What can you observe about gemstones?

_____
_____
_____
_____
_____

Activity 14
Analyze Like a Scientist

Quick Code: us2339s

Reverse Engineering

Read the text and **look** at the photo. Then, **think** about how you can learn how a kaleidoscope works.

Read Together

Reverse Engineering

People search the landscape for gemstones. Many gemstones are rare treasures. People like to look at them. Engineers have designed a toy that gives people a chance to enjoy these beautiful treasures. A kaleidoscope is a toy made from materials that look like gemstones.

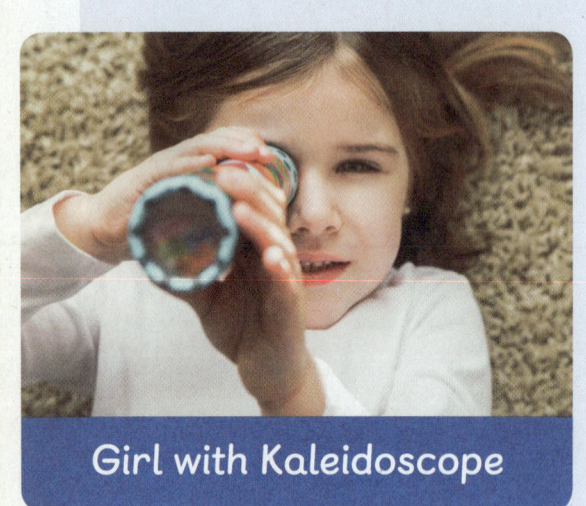

Girl with Kaleidoscope

Have you ever looked inside a kaleidoscope? You can see what look like tiny gemstones arranged in a pattern. How does a kaleidoscope work? You can find out by taking apart the object and looking at each part. How do the parts fit together? When you take apart an object to find out how it works and what it is made of, this is called reverse engineering.

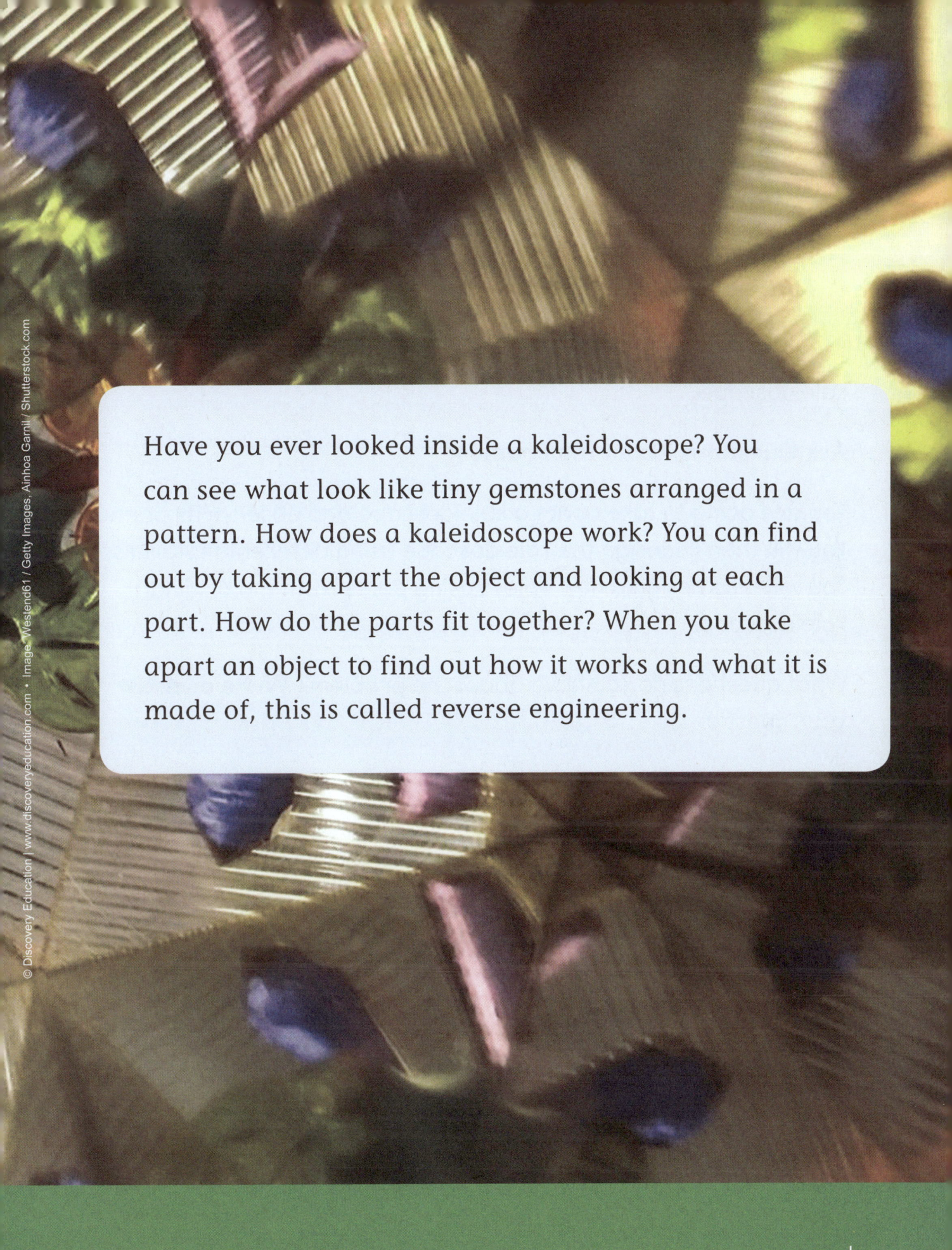

2.3 | Learn — How are materials chosen for designs?

Activity 15
Design Solutions Like a Scientist

Quick Code: us2340s

Hands-On Engineering: Designing a Kaleidoscope

In this activity, you will engage in reverse engineering to design a kaleidoscope.

Ask Questions About the Problem

You are going to take apart a kaleidoscope and observe its parts. You will then redesign the kaleidoscope using different materials. You will also design a new object using parts from the original kaleidoscope.

What questions do you have about the problem? **Write** or **draw** your questions.

SEP Planning and Carrying Out Investigations
CCC Structure and Function
CCC Patterns

What materials do you need? (per group)

- Kaleidoscope
- Beads
- Shiny, mirrored paper
- Sequins
- Scissors
- Cellophane
- Construction paper
- Wrapping paper sheets
- Clear tape
- Glue
- String
- Markers
- Stickers
- Paper confetti in different shapes (optional)
- Plastic zipper bags (optional)

What Will You Do?

Take apart the kaleidoscope. **Draw** a picture of its parts. **Label** each part. Tell what each part does. If you need more space, draw on another sheet of paper.

2.3: Materials in Design

2.3 | Learn   How are materials chosen for designs?

Think about how you will redesign the kaleidoscope using different parts. How do you think the different parts will change the way the kaleidoscope works?

_____

_____

_____

Now redesign your kaleidoscope. How does this kaleidoscope work? **Write** or **draw**.

Use parts from the original kaleidoscope to design a new useful object. What did you design? How does this kaleidoscope work? **Write** or **draw**.

2.3: Materials in Design

2.3 | Learn   How are materials chosen for designs?

Think About the Activity

What did each part of the kaleidoscope do? What were the strengths of your redesign? What were the weaknesses of your redesign? **Write** your answers in the chart.

Part of the Kaleidoscope	
Strengths of redesign:	Weaknesses of redesign:

Activity 16
Observe Like a Scientist

Quick Code: us2341s

How Does a Kaleidoscope Work?

Watch the video. Then, **talk** about what you observe.

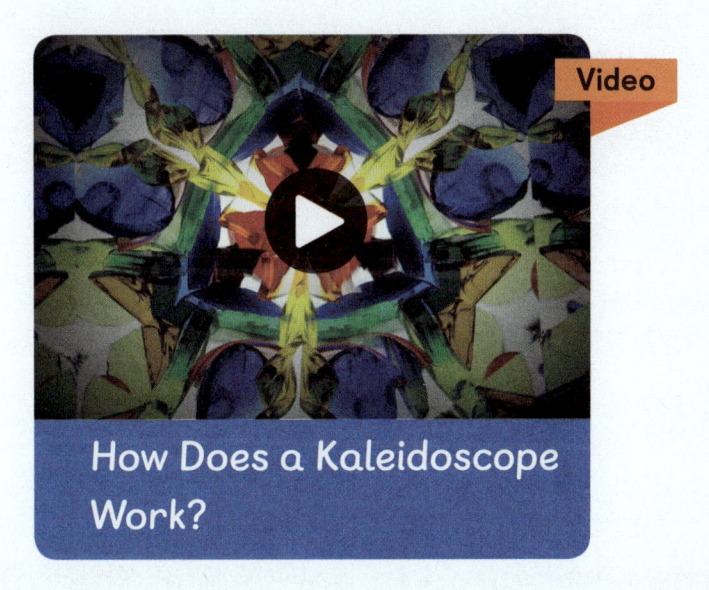

Video

How Does a Kaleidoscope Work?

Talk Together

Now, talk together about how a kaleidoscope works.

2.3: Materials in Design | 191

Activity 17
Analyze Like a Scientist

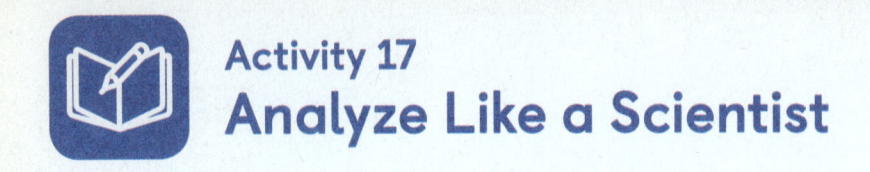

Quick Code: us2342s

Parts Working Together

Read about the parts of a bicycle.

> **Read Together**
>
> # Parts Working Together
>
> Ravi is Padma's older brother. Ravi is taking apart a bicycle, and Padma is helping. Her job is to store each part carefully so it will not be lost.
>
> First, Ravi takes off the wheels. Padma decides to put all the parts that are round in one place. She rolls the wheels to the side of the garage.
>
> Parts Working Together
>
> Next, Ravi takes off the seat. The seat is shaped like a triangle. This shape makes the seat comfortable to sit on. Padma puts it in another spot.

Then, Ravi takes off the pedals. The pedals are shaped like rectangles. A rectangle shape makes it easy for feet to push on the pedals.

Next off are the gear and chains. The gear is round and made of metal. It has teeth around its edges. The chain is long. The holes in the chain catch the teeth of the gear. The gear and chain work together to connect the pedals to the wheels.

Finally, all that remains are the frame and the handlebars. Padma tells Ravi that the parts no longer make a bicycle. Ravi tells her not to worry.

"After I put the parts back together, " he says, "they will make a bicycle again!"

Handlebars on a Bike

Draw each part of the bicycle. **Label** each part.

Activity 18
Observe Like a Scientist

Quick Code: us2343s

Using Materials: Building Structures

Watch the video. **Look** for ways that straws were used to make straw towers. Then, follow your teacher's directions to **draw** a sketch of a straw tower. Finally, **talk** about your design.

Video

Using Materials: Building Structures

Talk Together

Now, talk together about your straw tower design to get feedback.

SEP Planning and Carrying Out Investigations

2.3: Materials in Design

2.3 | Learn How are materials chosen for designs?

Activity 19
Observe Like a Scientist

Quick Code: us2344s

Inside a Kaleidoscope

Look at the photo. **Answer** the questions.

Inside a Kaleidoscope

What shapes and designs do you see in this kaleidoscope?

_____

_____

How are new designs created in a kaleidoscope?

_____

_____

_____

SEP Engaging in Argument from Evidence

Activity 20
Evaluate Like a Scientist

Quick Code: us2345s

Make It Better

Think about how you could improve the shape or function of the chair or desk you use at school. **Write** or **draw** what you would like to change. **Tell** why you would want to change it. Then, make a drawing that shows what you would do to improve the design of your chair or desk.

SEP Engaging in Argument from Evidence

2.3: Materials in Design

2.3 | Learn How are materials chosen for designs?

How Can Engineers Be Creative When Using Materials?

Activity 21
Observe Like a Scientist

Quick Code: us2346s

Landfills

Look at the photo. **Think** about how the pile of trash was made. Then **talk** about what you observe.

Landfills

Talk Together

Now, talk together about how garbage is taken to a landfill.

198

Activity 22
Observe Like a Scientist

Quick Code: us2347s

Garbage

Watch the video. Then, **talk** about what you observe.

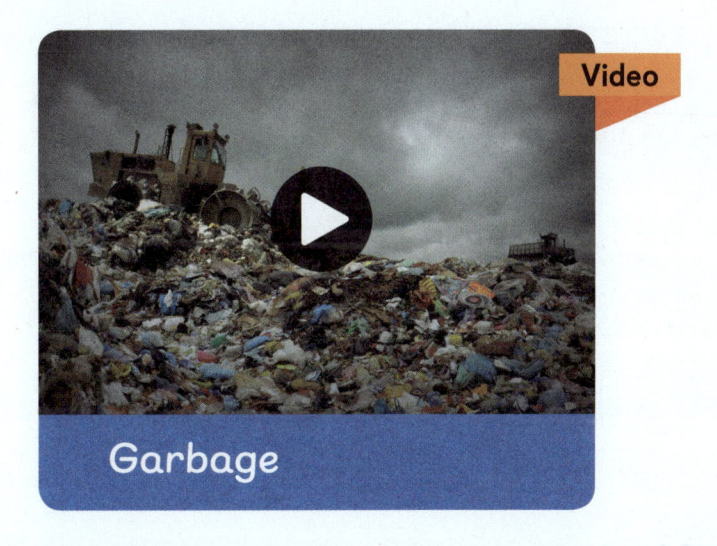

Video

Garbage

Talk Together

Now, talk together about what is true and not true about trash that is thrown away.

2.3: Materials in Design | 199

Activity 23
Analyze Like a Scientist

Quick Code:
us2348s

Using Materials Wisely

Read the text about using materials wisely.

> 🧑‍🏫 **Read Together**
>
> # Using Materials Wisely
>
> Humans have the ability to shape the land. One way they do this is by burying trash in **landfills**.
>
> When trash is thrown away, it can end up in a landfill. Over time, some materials in the landfill break down easily. These materials are biodegradable. Other objects do not break down easily. If too many objects that do not break down easily are placed in the landfill, the landfill will fill up too quickly. We will run out of places to put the trash.
>
> Landfill

200

Engineers can help protect the land. Using recycled materials in products stops many items like plastics from filling up landfills. Engineers think creatively about how to use recycled and reused products in their designs.

Recycling Plant Engineer

Why are many landfills filling up too quickly?

Activity 24
Observe Like a Scientist

Quick Code: us2349s

Science Lab

Watch the video. **Look** for evidence about types of materials that break down easily.

Science Lab

Talk Together

Now, talk together about the evidence about types of materials that break down easily.

SEP Obtaining, Evaluating, and Communicating Information

2.3: Materials in Design | 203

2.3 | Learn How are materials chosen for designs?

Activity 25
Observe Like a Scientist

Quick Code: us2350s

California Cleaning

Watch the video. **Look** for problems that trash can cause on the beach.

California Cleaning

Talk Together

Now, talk together about possible solutions to too much trash on the beach.

Activity 26
Observe Like a Scientist

Quick Code: us2351s

Reusing Materials

Watch the video. **Look** for plastic and metal objects.

Reusing Materials

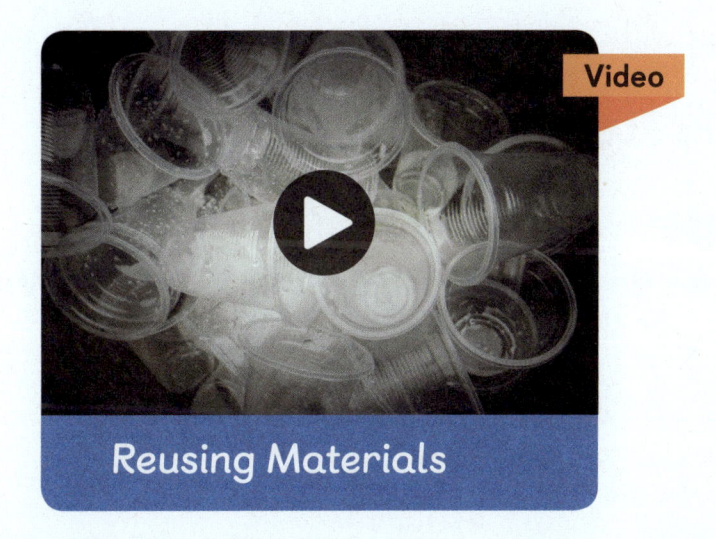

Talk Together

Now, talk together about the uses for plastic and metal objects.

2.3: Materials in Design | 205

2.3 | Learn — How are materials chosen for designs?

Activity 27
Design Solutions Like a Scientist

Quick Code: us2352s

Hands-On Engineering: The Right Stuff

In this activity, you will explore the properties of different materials. You will use the materials to make different things.

Ask Questions About the Problem

You are going to use materials to make different useful objects. What questions do you have about the problem? **Write** or **draw** your questions.

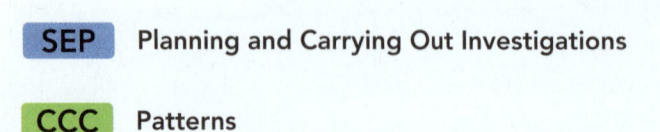

SEP Planning and Carrying Out Investigations

CCC Patterns

What materials do you need? (per group)

- 2 matching felt rectangles, with holes punched along the side
- Metal brads, 6
- Aluminum foil
- Cardboard paper towel tube
- Cotton balls, small bag
- Craft sticks
- Glue
- Tissue paper
- Construction paper
- Crayons
- Yarn
- Clear tape
- Scissors
- Metric ruler
- Empty plastic soda/water bottles from home (washed)
- Empty soda cans from home (washed)
- Plastic toy building block

2.3 | Learn — How are materials chosen for designs?

What Will You Do?

Look at the materials you can use. **Look** at their properties. Decide what you want to make. **Write** or **draw** your ideas.

Think About the Activity

Write or **draw** to complete the chart.

Name Three Materials	Describe Their Properties

How Did You Use These Materials? Why?

2.3: Materials in Design

2.3 | Learn — How are materials chosen for designs?

Activity 28
Evaluate Like a Scientist

Quick Code: us2353s

Reuse It

You can save money and help the environment by reusing materials. **Look** at each object. How can you reuse it? For each object, **describe** three new ways it can be used that are different from its normal use.

Tin Cans

SEP Engaging in Argument From Evidence

Socks

Plastic Bottle

2.3: Materials in Design | 211

2.3 | Share How are materials chosen for designs?

Activity 29
Record Evidence Like a Scientist

Quick Code: us2354s

Three Pigs Problem

Now that you have learned about materials in design, look again at the Three Pigs Problem. You first saw this in Wonder.

Let's Investigate Three Pigs Problem (Video)

Talk Together

How can you describe the Three Pigs Problem now? How is your explanation different from before?

SEP Constructing Explanations and Designing Solutions

Look at the Can You Explain? question. You first read this question at the beginning of the lesson.

> ## Can You Explain?
> How are materials chosen for designs?

Now, you will use your new ideas about the Three Pigs Problem to answer a question.

1. **Choose** a question. You can use the Can You Explain? question, or one of your own. You can also use one of the questions that you wrote at the beginning of the lesson.

   My Question

2. Then, use the sentence starters on the next page to **answer** the question.

2.3: Materials in Design | 213

2.3 | Share — How are materials chosen for designs?

The materials in a design are chosen because
_____
_____
_____

Based on my observations, I think
_____

is a good way to come up with design solutions because
_____
_____
_____

An example of reverse engineering would be
_____
_____
_____

The evidence I collected shows
_____
_____
_____
_____

STEM in Action

Quick Code: us2355s

Activity 30
Analyze Like a Scientist

Sorting Trash

Read the text. Then **talk** about recycling.

Read Together

Sorting Trash

Think about the **matter** you throw away every day. Where does it go? In the garbage can? In the recycling bin? What about a compost pile? Some things are put down the drain.

How do you decide? You probably think about what the objects are made of. Metal, glass, and paper can be recycled to make new things. Some kinds of plastics can also be recycled.

Recycling

SEP Obtaining, Evaluating, and Communicating Information

2.3: Materials in Design | 215

Read Together

You may choose to throw items in the garbage. You may also recycle. Recycled objects are sorted by material.

People at recycling centers think about what objects are made of. They sort the objects by type. Glass goes in one place. Metal goes in another. Plastic and paper go in different areas. Each type is recycled separately.

Video

How to Recycle

Other people find ways to use recycled objects. They observe what the objects are made of. Metal cans can be melted. They form new metal cans. Plastic bottles can be shredded. The plastic is used to make park benches.

Conservation of Resources

There are many reasons to conserve our **resources**. One of the most important reasons is that conserving resources helps to protect living things. But what does conservation mean?

Conservation is the careful use of Earth's natural resources. By not using too much of Earth's resources, we can help protect the environment.

One way to conserve resources is by saving something you have already used. For example, you can use scrap paper left over from math assignments for art projects.

We practice conservation by reducing, reusing, and recycling materials.

2.3: Materials in Design

Read Together

Reusable Lunch Bags

You can also reuse a resource. For example, a cloth lunch bag can be used over and over again. Instead of using a paper lunch bag that you throw out after each lunch, you can use a cloth bag every day.

We can also protect the environment by using as little of a resource as possible. When we reduce the amount of a resource we use, this means that we do not need to take as much of it from the environment. For example, if you do not waste paper, fewer trees will be needed to make more paper.

Recycling helps to conserve resources.

Many of the resources we use can be recycled. This means that they can be collected and then turned into new materials that can be used again. For example, recycling plastic bottles helps to make other things that are made out of plastic.

Many different types of materials can be recycled. These include metals, plastic, paper, and glass. Recycling bins are labeled with symbols indicating what materials go where. You should sort recyclable materials by these labels.

| Recycle | Plastic | Glass | Cans | Paper | Cardboard |

Many schools and communities are making an effort to recycle more and throw out less. Food and plant waste can be composted. Old clothes can be used for bedding in animal shelters. Rain can be caught and stored in order to water plants. Some people place gravel on their lawns instead of grass so they do not have to water and fertilize it. By recycling as much as we can, we reduce the amount of what we throw out.

Recycling Bins

You are putting objects in recycling bins. Which bin should you use? Write **Plastic**, **Glass**, **Metal**, or **Paper** below each object.

1 pound	2 pounds	5 pounds
_____	_____	_____
1 pound	5 pounds	1 pound
_____	_____	_____

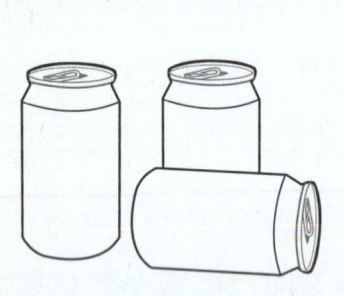

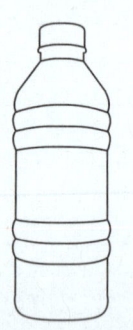

Create a bar graph of the recycling data. **Add** to find out the total number of pounds recycled.

Recycling Bins

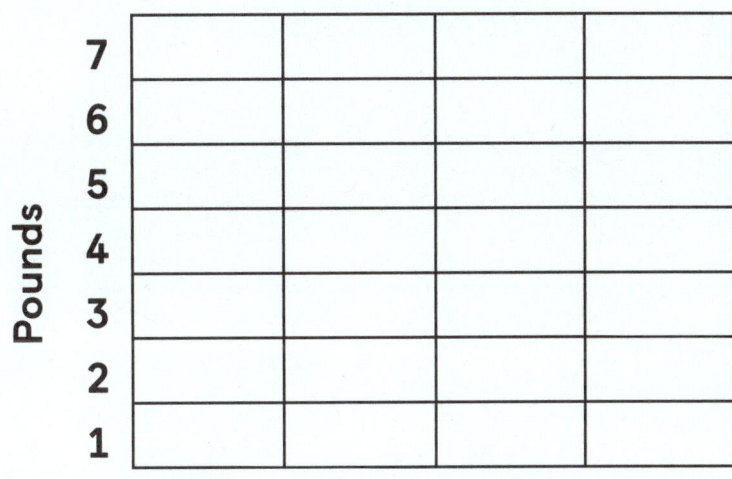

	Plastic	Glass	Metal	Paper
7				
6				
5				
4				
3				
2				
1				

Pounds

How many pounds did you recycle altogether?

_____

2.3: Materials in Design

2.3 | Share How are materials chosen for designs?

Activity 31
Evaluate Like a Scientist

Quick Code: us2356s

Review: Materials in Design

Think about what you have read and seen.
What did you learn?
Draw what you have learned.
Then, **tell** someone else about what you learned.

Talk Together

Think about what you saw in Get Started. Use your new ideas to discuss materials in design.

CCC Patterns

2.3: Materials in Design

Unit Project

Design Solutions Like a Scientist

Quick Code: us2357s

Hands-On Engineering: Choosing the Best Materials

In this activity, you will take apart an object. You will observe how its parts work together. Then you will design an object that performs a similar function using recycled or reused materials. You will compare the strengths of different materials and describe how those materials are best suited for their purpose.

Organized Recycling Bin

SEP Constructing Explanations and Designing Solutions
CCC Patterns
CCC Structure and Function

What materials do you need? (per group)
- Toy trucks
- Screwdriver set
- Safety goggles (per student)
- Various materials that are recycled or reused
- Battery fan
- Pliers
- Wrench

HANDS-ON ENGINEERING

What Will You Do?

Look at the recycling bin. What is the problem?

_____

_____

What ideas do you have to solve the problem? **Draw** and **write** about it.

Unit 2: Materials from the Land | 225

Unit Project

How will you pick one idea to test?

_____
_____
_____
_____

How will you know your idea works?

_____
_____

Test your design. **Draw** or **write** to show how you tested it.

Think About the Activity

Write or **draw** your answers to the questions in the chart. How well did your design perform a similar function to a recycling bin? How could you improve your design?

What Worked?	What Did Not Work?

What Could Work Better?

Unit 2: Materials from the Land

Grade 2 Resources

- Bubble Map
- Safety in the Science Classroom
- Vocabulary Flash Cards
- Glossary
- Index

Name _____

Bubble Map

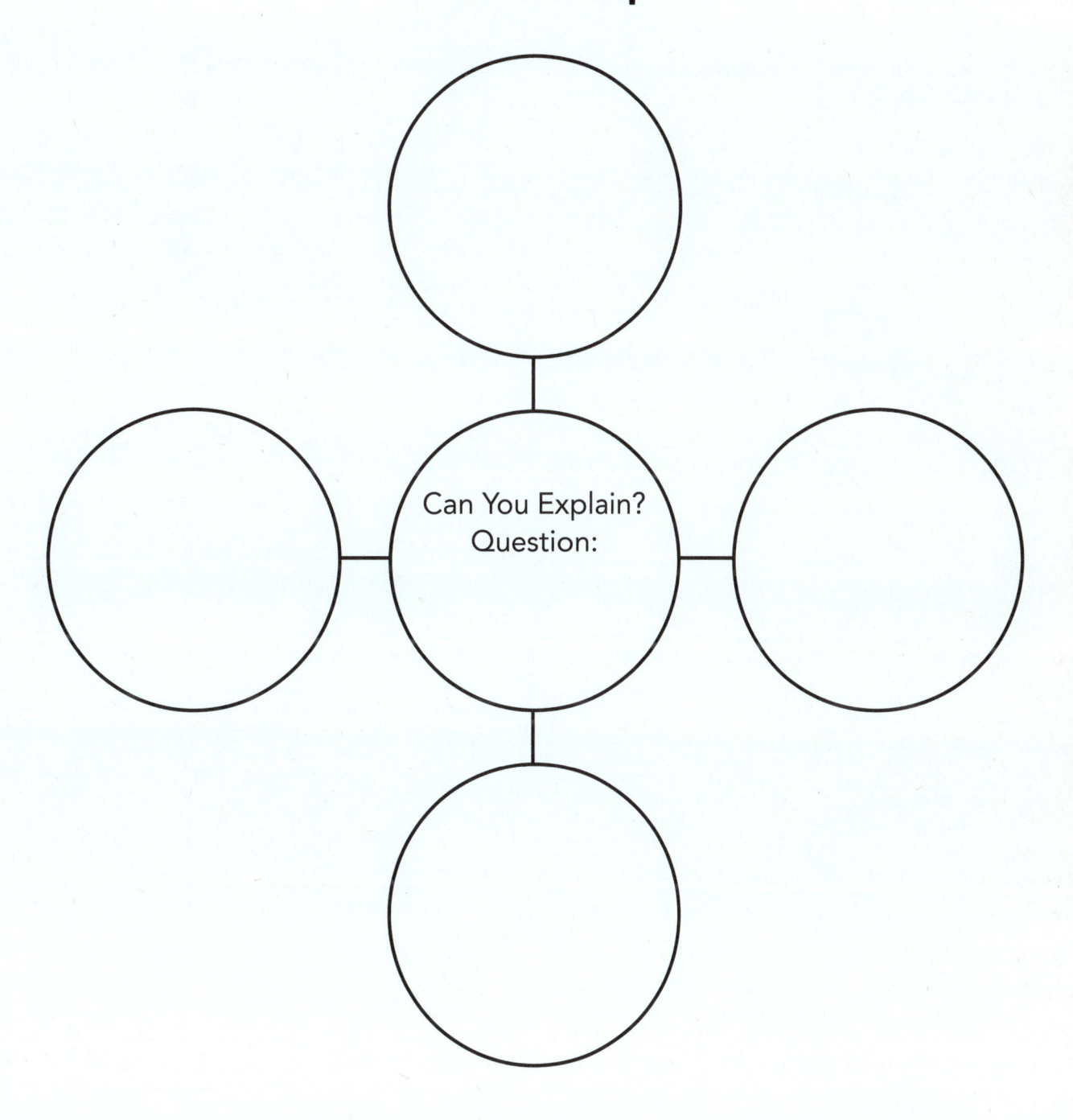

Can You Explain?
Question:

Bubble Map | R1

Safety in the Science Classroom

Following common safety practices is the first rule of any laboratory or field scientific investigation.

Dress for Safety

One of the most important steps in a safe investigation is dressing appropriately.

- Splash goggles need to be kept on during the entire investigation.

- Use gloves to protect your hands when handling chemicals or organisms.

- Tie back long hair to prevent it from coming in contact with chemicals or a heat source.

- Wear proper clothing and clothing protection. Roll up long sleeves, and if they are available, wear a lab coat or apron over your clothes. Always wear closed-toe shoes. During field investigations, wear long pants and long sleeves.

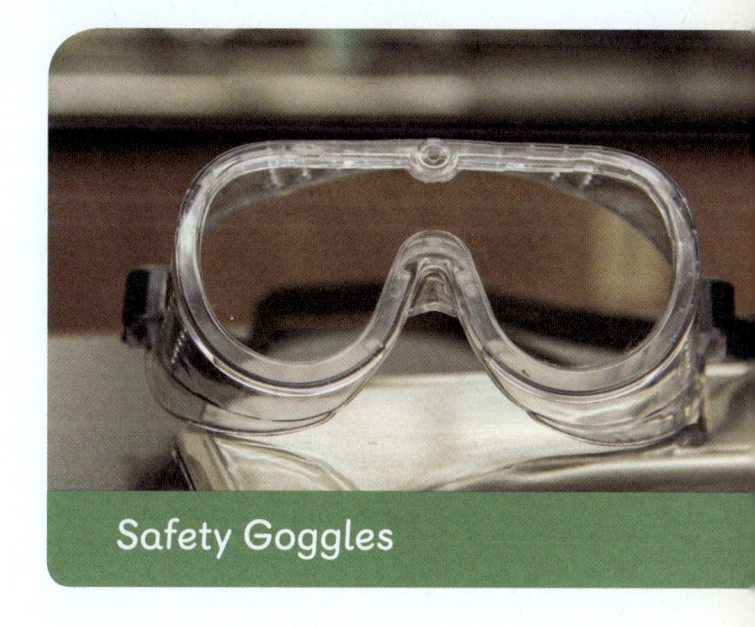

Safety Goggles

Safety in the Science Classroom | R3

Be Prepared for Accidents

Even if you are practicing safe behavior during an investigation, accidents can happen. Learn the emergency equipment location in your classroom and how to use it.

- The eye and face wash station can help if a harmful substance or foreign object gets into your eyes or onto your face.

- Fire blankets and fire extinguishers can be used to smother and put out fires in the laboratory. Talk to your teacher about fire safety in the lab. He or she may not want you to directly handle the fire blanket and fire extinguisher. However, you should still know where these items are in case the teacher asks you to retrieve them.

Most importantly, when an accident occurs, immediately alert your teacher and classmates. Do not try to keep the accident a secret or respond to it by yourself. Your teacher and classmates can help you.

Fire Extinguisher

Practice Safe Behavior

There are many ways to stay safe during a scientific investigation. You should always use safe and appropriate behavior before, during, and after your investigation.

- Read all of the steps of the procedure before beginning your investigation. Make sure you understand all the steps. Ask your teacher for help if you do not understand any part of the procedure.

- Gather all your materials and keep your workstation neat and organized. Label any chemicals you are using.

- During the investigation, be sure to follow the steps of the procedure exactly. Use only directions and materials that have been approved by your teacher.

- Eating and drinking are not allowed during an investigation. If asked to observe the odor of a substance, do so using the correct procedure known as wafting, in which you cup your hand over the container holding the substance and gently wave enough air toward your face to make sense of the smell.

- When performing investigations, stay focused on the steps of the procedure and your behavior during the investigation. During investigations, there are many materials and equipment that can cause injuries.

- Treat animals and plants with respect during an investigation.

- After the investigation is over, appropriately dispose of any chemicals or other materials that you have used. Ask your teacher if you are unsure of how to dispose of anything.

- Make sure that you have returned any extra materials and pieces of equipment to the correct storage space.

- Leave your workstation clean and neat. Wash your hands thoroughly.

Vocabulary Flash Cards

absorb
to take in or soak up

absorption
how much something can take in and hold

dissolve
to mix something with a liquid such as water so that it can't be seen anymore

engineer
a person who designs something that may be helpful to solve a problem

Vocabulary Flash Cards | R7

flexibility

the ability to bend without breaking

gemstone

a colorful stone found in nature that can be used for jewelry

hardness

a measure of how difficult it is to scratch a mineral: Diamonds are the hardest mineral. They have a hardness scale rating of 10.

landfill

a place where trash is buried

Vocabulary Flash Cards | R9

landscape

the view of a land's surface

material

things that can be used to build or create something

matter

the things around you that take up space like solids, liquids, and gases

mixture

a combination of different things, but you can pick out each different one

Vocabulary Flash Cards | R11

property

a characteristic of something

resource

a material that can be used to solve problems

reverse engineering

the process of learning about something by taking it apart to see how it works, and what it is made of

soil

dirt that covers Earth, in which plants can grow and insects can live

Vocabulary Flash Cards | R13

solution

a combination of two things that are mixed so well that each one cannot be picked out

Vocabulary Flash Cards | R15

Glossary

English ——— A ——— Español

absorb
to take in or soak up

absorber
tomar o captar

absorption
how much something can take in and hold

absorción
cuanto algo puede tomar y retener

adjust
to fix or change something

ajustar
arreglar o cambiar algo

analyze
to closely examine something and then explain it

analizar
examinar con atención algo y luego explicarlo

——— B ———

barrier
something that is used to stop or block materials from moving

barrera
algo que se usa para evitar o bloquear el movimiento de materiales

biodiversity
the many different types of life that live together in an environment

biodiversidad
muchos y diferentes tipos de vida que conviven en un medio ambiente

--- C ---

canyon
a deep valley that has very steep sides

cañón
valle profundo que tiene laderas muy pronunciadas

channel
a path that is dug and used for drainage or protection against things like water, mud, or rocks

canal
vía cavada que se usa como desagüe o protección contra cosas como el agua, el lodo o las rocas

characteristic
a special quality that something may have

característica
cualidad especial que tiene algo

--- D ---

dissolve
to mix something with a liquid, such as water, so that it can't be seen anymore

disolver
mezclar algo con un líquido, como el agua, de manera que no se pueda ver más

drought
when there is no rain for a long period

sequía
cuando no llueve durante un período prolongado

E

Earth's crust
the top layer of Earth that is the thinnest and the most important because it is where we live

corteza de la Tierra
capa superior de la Tierra que es la más delgada y la más importante porque allí es donde vivimos

earthquake
a sudden shaking of the ground caused by the movement of rock underground

terremoto
repentina sacudida de la tierra causada por el movimiento de roca subterránea

elevation
the height of an area of land above sea level

elevación
altura de un área de tierra por encima del nivel del mar

engineer
a person who designs something that may be helpful to solve a problem

ingeniero
persona que diseña algo que puede ser útil para resolver un problema

engineering
using math and science to design and build machines, structures, and other devices

ingenería
usar las matemáticas y las ciencias para diseñar y construir máquinas, estructuras y otros dispositivos

environment
all the living and nonliving things that surround an organism

medio ambiente
todos los seres vivos y objetos sin vida que rodean a un organismo

erosion
when soil is moved from one location to another by wind or water

erosión
cuando el viento o el agua transporta suelo de un lugar a otro

estimate
to make a careful guess

estimar
hacer una suposición consciente

—— F ——

feature
a thing that describes what something looks like; part of something

rasgo
cosa que describe cómo se ve algo; parte de algo

flexibility
the ability to bend without breaking

flexibilidad
capacidad de doblarse sin romperse

fresh water
water that is not salty, such as that found in streams and lakes

agua dulce
agua que no es salada, como la que se encuentra en arroyos y lagos

G

gemstone
a colorful stone found in nature that can be used for jewelry

piedra preciosa
piedra colorida que se encuentra en la naturaleza y se puede usar para hacer joyas

H

habitat
the place where a plant or animal lives

hábitat
lugar donde vive una planta o un animal

hardness
a measure of how difficult it is to scratch a mineral: Diamonds are the hardest mineral. They have a hardness scale rating of 10.

dureza
medida de cuán difícil es rayar un material: los diamantes son los minerales más duros. Su clasificación en la escala de dureza es 10.

L

landfill
a place where trash is buried

vertedero
lugar donde se entierra la basura

landform
a feature of Earth that has been formed by nature, such as a hill or a valley

accidente geográfico
característica de la Tierra formada por la naturaleza, como una colina o un valle

landscape
the view of a land's surface

paisaje
vista de la superficie de un terreno

location
a place where something is

ubicación
lugar donde se encuentra algo

M

map
a flat picture or drawing of a place that is made to show things, such as streets or towns, in an area

mapa
imagen o dibujo plano de un lugar que se hace para mostrar cosas, como las calles o las ciudades, de un área

material
things that can be used to build or create something

material
cosas que se pueden usar para construir o crear algo

matter
the things around you that take up space like solids, liquids, and gases

materia
cosas que nos rodean y ocupan espacio, como los sólidos, los líquidos y los gases

mixture
a combination of different things, but you can pick out each different one

mezcla
combinación de diferentes cosas, pero se puede identificar cada una

model
a human-made version created to show the parts of something else, either big or small

modelo
versión creada por el hombre para mostrar las partes de algo más, ya sea grande o pequeño

mountain
a very tall area of land that is higher than a hill and has steep sides

montaña
área de tierra muy alta que es más alta que una colina y tiene laderas pronunciadas

N

naturalist
someone who studies nature, especially plants and animals

naturalista
alguien que estudia la naturaleza, especialmente las plantas y los animales

nutrient
something in food that helps people, animals, and plants live and grow

nutriente
algo en los alimentos que ayuda a las personas, los animales y las plantas a vivir y crecer

O

observe
to watch closely

observar
mirar atentamente

ocean
a large body of salt water

océano
gran cuerpo de agua salada

organism
a living thing

organismo
ser vivo

P

plain
a large flat area of land without trees

llanura
gran área de tierra llana sin árboles

plateau
a large, flat area of land that is higher than the other land around it

meseta
gran área de tierra llana que está a más altura que el terreno que la rodea

pollen
the yellow powder found inside a flower

polen
polvo amarillo que se encuentra dentro de una flor

pollination
moving or carrying pollen from a plant to make the seeds grow

polinización
transferencia o transporte de polen de una planta para hacer que crezcan las semillas

preserve
to protect or keep something safe

preservar
proteger o mantener algo a salvo

property
a characteristic of something

propiedad
característica de algo

--- Q ---

quadrilateral
a flat shape with four straight sides, such as a square or a parallelogram

cuadrilátero
figura plana con cuatro lados rectos, como un cuadrado o un paralelogramo

R

recycle
to create new materials from something already used

reciclar
crear nuevos materiales a partir de algo usado

relief map
a type of map that shows how flat or steep the landforms are in an area

mapa de relieve
tipo de mapa que muestra si los accidentes geográficos son llanos o pronunciados en un área

resource
a material that can be used to solve problems

recurso
material que se puede usar para resolver problemas

restore
to put into use again

restablecer
volver a poner en servicio

reverse engineering
the process of learning about something by taking it apart to see how it works and what it is made of

ingeniería inversa
proceso de aprender acerca de algo, desarmándolo para ver cómo funciona y de qué está hecho

river
water flowing through a landscape, usually fed by smaller streams

río
agua que fluye a través de un área, por lo general alimentada por arroyos más pequeños

--- S ---

select
to choose or pick

seleccionar
elegir o escoger

shelter
a place that protects you from harm or bad weather

refugio
lugar para protegerse de peligros o el mal tiempo

slope
land that is slanted or angles downward

pendiente
tierra inclinada hacia abajo

soil
dirt that covers Earth, in which plants can grow and insects can live

suelo
tierra que cubre nuestro planeta en la que pueden crecer plantas y vivir insectos

solution
a combination of two things that are mixed so well that each one cannot be picked out

solución
combinación de dos cosas que se mezclan tan bien que no se puede identificar cada una

strategy
a plan that can solve a problem

estrategia
plan que puede resolver un problema

stream
a small flowing body of water that starts with a spring and ends at a river

arroyo
pequeño cuerpo de agua que fluye y nace en una vertiente y termina en un río

survive
to continue to live

sobrevivir
continuar viviendo

— T —

two-dimensional
drawings and sketches that are done on flat paper to show width and height

bidimensional
dibujos y bosquejos que se hacen en papel plano para mostrar el ancho y la altura

V

valley
the low place between two hills or mountains

valle
lugar bajo entre dos colinas o montañas

W

weathering
the breakdown of rocks into smaller pieces called sediment

meteorización
desintegración de rocas en trozos más pequeños llamados sedimento

Index

A

Absorption 36, 54–60
Adobe 122, 123–124, 125
Analyze Like a Scientist 12, 32–33, 36, 51, 65–67, 79, 88–89, 96–99, 101–102, 116–119, 121–122, 133, 142–144, 152–154, 158–159, 162–163, 168–169, 178–179, 182–183, 190–192, 198–200, 213–214
Ask Questions Like a Scientist 10–11, 74–75, 150–151
Asphalt
    landscapes and 32–33
    steps involved in making 27–29

B

Beach 202
Beans 103–107
Bicycle 190–192
Biodegradable materials 198
Bottles, plastic 214
Bricks
    adobe 121–122, 123
    building with 74–78, 79
    changing 79
    making 76–77
    mixtures and 98
Building materials. *See also specific materials*
    mixtures and 92–93

C

Can You Explain? 8–9, 72, 148
Chairs 195
Chalk rock 34–35
Change
    adobe and 120, 121–122, 123
    bricks and 79
    of clay for job 142–144
    of clay shape 128–130
    clay temperature and 131, 132
    describing 137, 145
    evaporation and 115
    layers and 110–112
    making bricks and 76–77
    mixed vs. separated 119
    mixing building materials and 92–93
    mixing materials together and 82, 88–89, 96–108, 100
    mixing solids and liquids and 83–87

recycling, sorting and 108
reversible vs. irreversible 72, 133, 134–135
rocks as mixtures and 94–95
separating mixtures and 101–102, 103–107, 109, 113–114, 116–118
shape, color, temperature and 80–82
solutions and 90–91
temperature and 120, 121–122, 123, 124–127, 131, 132
Three Little Pigs and 138–141
Classifying classroom objects 43–44
Clay
changing 128–130
job changing 142–144
reversible and irreversible changes to 134–135
soil and 30–31, 66
temperature and 131, 132
Color, change and 80–82
Comparing
classroom objects 43–44
materials 50, 51, 54–60

Composting 217
Concrete
landscapes and 32–33
mixtures and 98
in schoolyard landscape 26
Conservation 215–219
Consumption, reducing 216
Crayons 179
Criteria 50

D

Design
bicycle parts and 190–192
choosing building materials and 148, 150–151, 152–154, 162–163
improving shape or function of 195
kaleidoscopes and 182–183, 184–188, 189
landfills and 196, 197, 198–200
materials and systems and 158–159
materials in kitchen and 160–161
recycling, reusing and 222–225

Index | R31

Design (cont.)
    reusing materials and 208–209
    reverse engineering and 155–157, 182–183
    of straw towers 193
    stucco in straw and 170–171
    testing material properties and 168–169
    using different materials 204–207
Design Solutions Like a Scientist 4–5, 155–157, 184–188, 204–207, 222–225
Desks, improving shape or function of 195
Dissolving 90–91
Drying of clay 131

E

Engineers
    comparing materials 50
    reverse engineering and 155–157
    selection of materials by 162–163
    soil scientists as 66
    testing material properties and 168–169
Evaluate Like a Scientist 20–21, 48–49, 61, 68–69, 119, 137, 145, 160–161, 195, 208–209, 220
Evaporation
    in Mojave Desert 115
    separating mixtures by 113–114

F

Filters 113–114, 117
Firing of clay 132
Flexibility 36
Flow, temperature and 124–127
Freezing 136
Function, improving 195

G

Garbage 197
Gases
    defined 36
    finding objects in state 48–49
Gemstones 180–181, 182–183
Glass in schoolyard landscape 25
Gravel 217

H

Hand lens, observing soil through 14–17
Hardness 36, 54–60
Hot chocolate 89
House
    bricks and 78
    materials used in 10–11, 62
    soil under 12, 67
Humus 66

I

Ice cubes 172–178
Investigate Like a Scientist 14–17, 38–42, 54–60, 83–87, 103–107, 110–112, 124–127, 128–130, 164–167, 172–178
Iron ore 28
Irreversible changes 72, 133, 134–136

J

Jackets 179

K

Kaleidoscopes 182–183, 184–188, 189, 194
Kitchen, materials in 160–161

L

Landfills 196, 197, 198–200
Landscapes
    different materials in 32–33
    mixtures and 98
    schoolyard 22–23, 24–26
    water in 46–47
Layers 110–112
Lemonade 97–98
Liquid asphalt 28
Liquids
    defined 36
    finding objects in state 48–49
    mixing of 83–87, 88–89
    solutions and 90–91
Loam 30–31
Lunch bags 216

M

Magnets, separating mixtures with 113–114, 117
Magnifying glass, observing soil through 14–17
Matter, describing 61
Melting
    preventing 172–178
    reversible vs. irreversible change and 136

Metal in schoolyard landscape 25–26
Minerals
   soil and 30–31
   used in transportation materials 28
Mittens 178
Mixtures
   building materials and 92–93, 98
   evaporation and 115
   examples of 100
   landscapes and 98
   layers and 110–112
   making and separating 113–114
   of materials 82
   recycling, sorting and 108
   rocks and 89, 94–95, 98
   separating 101–102, 103–107, 108, 109, 113–114, 116–119, 117, 119
   soil and 98
   of solids and liquids 83–87, 88–89
   with and without changing 96–99
Mojave Desert 115
Mud 101–102

O

Objects
   describing 36
   investigating and classifying 43–44
Observe Like a Scientist 18–19, 22–23, 27–29, 30–31, 34–35, 37, 43–44, 45, 46–47, 52, 53, 76–77, 78, 82, 90–91, 92–93, 94–95, 100, 108, 109, 113–114, 115, 120, 123, 131, 132, 136, 170–171, 180–181, 189, 193, 194, 196, 197, 201, 202, 203
Organic material 30–31

P

Pavement in schoolyard landscape 25
Plastic in schoolyard landscape 26
Potters 142–144
Product testing 53
Properties
   of asphalt and steel 27–29
   for building house 10–11
   of chalk rock 34–35

of classroom objects 43–44
comparing 8
comparing materials by 50, 51, 54–60
defined 36
describing 45
design improvement and 52
of different toys 20–21
engineers and 50
of landscape materials 32–33
of matter 61
physical states 36, 48–49
product testing and 53
of rocks 37, 38–42
of sand 18–19
of schoolyard landscape 22–23, 24–26
selection of materials and 10–11, 62–64
soil scientists and 65–67
of soil types 14–17, 30–31
of soil under house 12
sorting rocks and 38–42
of water in landscape 46–47

R

Record Evidence Like a Scientist 62, 138–141, 210–212

Recycling 108, 150–151, 199, 213–214
Recycling bins 217, 218
Resource conservation 215–219
Reusing materials 150–151, 203, 208–209, 215–216
Reverse engineering 155–157, 182–183
Reversible changes 72, 133, 134–136
Rice 103–107
Rocks
　chalk 34–35
　differences in 37
　mixtures and 89, 94–95, 98
　observing and sorting 38–42
　soil and 30–31

S

Salads 88–89, 98
Salt 90–91
Sand
　beach and 18–19
　separating mixtures with 103–107
　soil and 30–31, 66
Schoolyard, materials in landscape of 22–23, 24–26

Selection of materials
   by engineers 162–163
   properties and 148, 150–151
   reuse, recycling and 222–225
   by Three Little Pigs 10–11,
      62–64, 74, 150–151, 210–212
Separating mixtures
   assessing 119
   recycling, sorting and 108
   of sand, rice, and beans
      103–107
   of soil and water 101–102, 109
   strategies for 116–119
   tools and properties used
      during 113–114
Shapes
   changing 80–82, 128–130
   improving 195
Silt 66
Size, separating mixtures by
   113–114
Socks 209
Soil
   of beach path 18–19
   under house 12, 67
   kinds of 30–31
   landscapes and 32–33
   mixtures and 98
   observing different kinds of
      14–17
   properties of 66
   scientists studying 65–67
   separating mixtures with
      101–102, 109
   testing water-holding ability of
      164–167
Soil scientists 65–67
Solids
   defined 36
   finding objects in state 48–49
   mixing of 83–87, 88–89
   solutions and 90–91,
      97–98
Solutions 90–91, 97–98
Sorting
   objects 43–44
   recycling and 108
   rocks 38–42
   of trash 213–214
Standards 50
Steel 27–29
Strange objects 178–179
Straw
   as building material 152–154
   building towers with 193
   stucco in 170–171

Strength 54–60
Stucco 170–171
Sugar 90–91, 97
Sunglasses 179
Syrup 124–127
Systems and materials in a home 158–159

T

Temperature
    adobe and 120, 121–122, 123
    change and 80–82
    changing materials with 120, 122, 124–127
    clay and 131, 132
    flow of syrup and 124–127
    freezing, melting and 136
    reversible and irreversible changes and 136
    separating mixtures with 113–114, 117
Thermos 172–178
Think Like a Scientist 24–26
Three Little Pigs
    bricks and 79
    changing materials and 138–141
    choosing building materials and 150–151, 152–154, 210–212
    design improvement and 52
    materials in design and 152–154
    properties of building materials and 51
    selection of materials by 10–11, 62–64, 72, 150–151, 210–212
    using and changing materials 138–141
Tin cans 208
Tools, changing clay with 128–130
Towers, straw 193
Transportation materials 27–29
Trash. *See also* Landfills
    on beach 202
    shaping the land with 2
    sorting of 213–214

W

Water
    ability of soil to hold 164–167
    in a landscape 46–47
    layers settling out of 110–112

Water (*cont.*)
    reversible changes to 72
    separating mixtures with
        101–102, 109, 117
    states of 46–47, 48–49

Wood
    as building material 152–154
    landscapes and 32–33
    reverse engineering house
        made of 155–157
    in schoolyard landscape
        25–26